PIC Cookbook
for Virtual Instrumentation

Richard Grodzik

PIC Cookbook
for Virtual Instrumentation

Richard Grodzik

Elektor International Media BV
P.O. Box 11
6114 ZG Susteren
The Netherlands

All rights reserved. No part of this book may be reproduced in any material form, including photocopying, or storing in any medium by electronic means and whether or not transiently or incidentally to some other use of this publication, without the written permission of the copyright holder except in accordance with the provisions of the Copyright, Designs and Patents Act 1988 or under the terms of a licence issued by the Copyright Licensing Agency Ltd, 90 Tottenham Court Road, London, England W1P 9HE. Applications for the copyright holder's written permission to reproduce any part of this publication should be addressed to the publishers.

The publishers have used their best efforts in ensuring the correctness of the information contained in this book. They do not assume, and hereby disclaim, any liability to any party for any loss or damage caused by errors or omissions in this book, whether such errors or omissions result from negligence, accident or any other cause.

British Library Cataloguing in Publication Data
A catalogue record for this book is available from the British Library

ISBN 978-0-905705-84-2

Prepress production: Kontinu, Sittard
Design cover: Helfrich Ontwerpbureau, Deventer
First published in the United Kingdom 2010
Printed in the Netherlands by Wilco, Amersfoort
© Elektor International Media BV 2010

099031-UK

Table of Contents.

       Acknowledgement
1   Introduction- basics
1.1   Delphi programming basics
1.2   RS232 basics
1.3   USB basics
1.4   USB RS232 serial and parallel bridges
1.5   PIC's and their ability to communicate using onboard UART and software 'bit-bang'
1.6   Terminal emulation

2   Delphi basics. Introduction to virtual instrumentation
2.1   A simple virtual thermometer
2.2   Abacus virtual components
2.3   Virtual flashing LED
2.4   Delphi Tcolorbutton

3   VCP USB communications
3.1   Installing FTDI VCP drivers
3.2   RS232 connectors
3.3   Portcontroller Active X control
3.4   Case study VCP DS75 virtual digital and analogue thermometer

4   DLL USB communications
4.1   Installing FTDI's DLL driver
4.2   Case Study DLL LED test board
4.3   Simple DAC for 0-5 volt output
4.4   Delphi VI slider control panel for test board
4.5   Delphi VI button control panel for test board
4.6   Delphi VI control panel for LED test board
4.7   Delphi VI gauge for PICKIT 2 44 pin microchip demo board

5   Case study joystick controlled mouse
5.1   The FTDI mini-B UB232R module and Schematic
5.2   Installing the joystick mouse
5.3   Software description

6   Case study simple Digital storage oscilloscope
6.1 Virtual instrument digital oscilloscope
6.2 A Delphi Lambda tester

7   Case study Virtual Compass
7.1 The compass hardware
7.2 The compass PC software
7.3 Operating the Compass
7.4 Testing the Compass

8   Case study FFT audio frequency analyser
8.1 FFT Simulator
8.2 FFT with ADC and serial output
8.3 Virtual spectrum display

9   Delphi virtual component suppliers
9.1 GMS Active X components
9.2 SDL Delphi virtual components
9.3 Uniworks virtual components
9.4 TMS Instrument workshop
9.5 IOCOMP virtual instrument components
9.6 CST software virtual components

10 Appendix
     unit D2XXUnit
     unit PORTCONTROLLERMODLib_TLB

10.1 References
10.2 Index

PIC Cookbook for virtual Instrumentation
By Richard Grodzik

Acknowledgement

I would like to dedicate this Book to my Teachers- Bob Kneeshaw Senior Lecturer at the Radio College, Ashley Down, Bristol (1971) and to Peter Nurse, Senior Lecturer at the Bradford and Ilkley Community College (1981), who taught me electronics and introduced me to the microprocessor.

Richard Grodzik January 2009.

Preface

The Delphi programming environment together with many 3rd party virtual component suppliers is ideally suited for the production of virtual instruments for the PC. This, together with the availability of USB/RS232 bridges and external Microchip PIC based intelligent sensors enables the PC to be used for data acquisition with data displayed on numerous varieties of virtual indicators including gauges, meters and displays on the PC's screen. Conversely the PC can also be used as an output device control center, with virtual switches, sliders and knobs manipulated by a mouse or keyboard controlling external devices and displays via the PC's USB port.

Detailed case studies in this Book include a virtual compass displayed on the PC's screen, a virtual digital storage oscilloscope, virtual -50 to +125 degree C thermometer, and FFT sound analyser, a joystick mouse and many examples detailing virtual instrumentation Delphi components. Arizona's embedded microcontrollers – the PIC's are used in the projects and include PIC16F84A, PIC16C71, DSPIC30F6012A, PIC16F877, PIC12F629 and the PIC16F887. Much use is made of Microchip's 44 pin development board (a virtual instrument 'engine)', equipped with a PIC16F887 with an onboard potentiometer in conjunction with the PIC's ADC to simulate the generation of a variable voltage from a sensor/transducer, a UART to enable PC RS232 communications and a bank of 8 LED's to monitor received data is also equipped with an ISP connector to which the 'PICKIT 2' programmer may easily be connected.

Now that the PC's RS232 serial port has been largely superseded by the USB port, it has been increasingly difficult to port data from the PC for control purposes and to import data for acquisition purposes. The old RS232 interface was easy to understand - a simple asynchronous serial data structure consisting typically of a start bit, an 8 bit data byte followed by a stop bit. The speed of transmisssion (baud rate) defined by the time width of each bit. In the case of USB data communications, the protocol is complex and the data is transmitted as 'packets' at a very fast rate, typically 10 Mbits/second.

Given that there are still countless electronic products equipped with an RS232 interface still in existance, including the author's 'Keighley 2000' 6.5 digit multimeter, and that very few new PC's are equipped with the serial RS232 'COM' port, using the RS232 port is no longer an option and the USB port has to be exclusively utilised for serial PC communications.

Thanks to FTDI's, Prolific and 4D's USB/RS232 convertors which provide a transparent bridge between the PC's USB port and external RS232 devices, serial communications for PC virtual instruments is now within the reach of everybody , hugely simplifying the task of PC interfacing. This Book features four of FTDI's DLL/VCP products – the MM232R mini USB-Serial UART, the USB mini-B UB232R bridge, the EVAL232R FT232RL USB to RS232 evaluation module and finally the DLP-USB245M parallel USB adapter. In addition the VCP Prolific USB convertor together with Scientific Instruments Active X 'PortController' is used for 'COM Port' communications.

I hope that the reader enjoys constructing some of the projects, since complete schematic drawings are included, including the complete source code for each case study. All of the code examples (PIC and Delphi) together with the HEX dumps and executable files may be downloaded from the Elektor.com website.

1 Introduction-basics

The main components to achieve virtual instrumentation in this book are the following: a PC 'front end' program – in this case Delphi software to produce a virtual instrument, communications-a PC USB/RS232 communication link, and an intelligent PIC controlled I/O interface board. This chapter introduces the reader to Delphi programming, RS232 and USB communications, USB/RS232 bridges (converters) and the fundamentals of PIC communication using RS232 and the use of terminal emulators.

1.1 Delphi programming basics

Delphi version 7 used throughout this book is commonly know as 'Pascal for Windows' using ready drawn components to create a 'front end' for a virtual instrument.Delphi allows you to create GUI (Graphical User Interface) or Console (text-only) applications (programs) along with many other types. We will concern ourselves here with the common, modern, GUI application. Delphi does a lot of work for us – the programmer simply uses the mouse to click, drag, size and position graphical component parts to build each screen of the application. Each part (or element) can be passive (displaying text or graphics), or active –in the case of a virtual component,(responding to a user mouse or keyboard action). This is best illustrated with a very simple program. Creating a simple 'Hello World' program When you first run Delphi, it will prepare on screen a new graphical application. This comprises a number of windows, including the menu bar, a code editor, and the first screen (form) of our program. Do not worry about the editor window at the moment.
The form should look something like this:

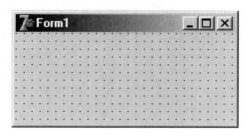

We have shown the form reduced in size for convenience here, but you will find it larger on your computer. It is a blank form, onto which we can add various controls and information. The menu window has a row of graphical items that you can add to the form. They are in tabbed groups : Standard, Additional, Win32 and so on. We will select the simplest from the Standard collection. Click on the **A** image to select a Label. This **A** will then show as selected:

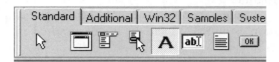

Having selected a graphical element, we then mark out on the form where we want to place the element. This is done by clicking and dragging. This gives us our first form element.:

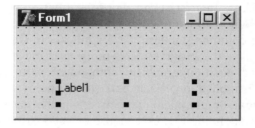

Notice that the graphical element contains the text **Label** as well as resize corners. The text is called the **Caption**, and will appear when we run the application. This **Caption** is called a **Property** of the button. The label has many other properties such as height and width, but for now, we are only concerned with the caption. Let us blank out the caption. We do this in the window called the **Object Inspector** (available under the **View** menu item if not already present:

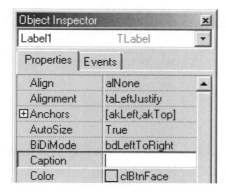

If we now return to the **Standard** graphical element collection, and select a button, shown as a very small button with **OK** on it, we can add this to the form as well:

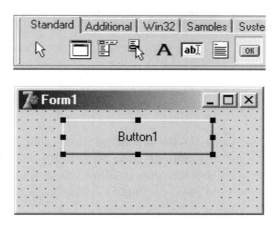

We now have a label and a button on the form. But the button will do nothing when pressed until we tell Delphi what we want it to do. So we must set an action, called an **Event**, for the button. The main event for a button is a **Click**. This can be activated simply by double clicking the button on the form.

This will automatically add an event called **OnClick** for the button, and add a related event handler in the program code:

```
procedure TForm1.Button1Click(Sender: TObject);
begin

end;
```

This 'skeleton' code will not do anything as it stands. We must add some code. Code that we add will run when the button is clicked. So let us change the label caption when the button is pressed. As we type, Delphi helps us with a list of possible options for the item we are working on. In our instance, we are setting a Label caption:

```
procedure TForm1.Button1Click(Sender: TObject);
begin
   Label1.Caption := 'Hello World';
end;
```

We assign a text value **'Hello World'** to the caption property. Note that we terminate this line of code with a **;** - all Delphi code statements end with this indicator. It allows us to write a command spread across multiple lines – telling Delphi when we have finished the command.

And we have now finished our very simple action – we will set the label to 'Hello World' when the button is pressed. To run the program, we can click on the Green triangle (like a Video play button), or press **F9**. When the program runs it looks like this:

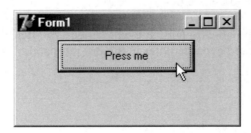

When we click on the button, we get:

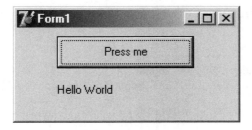

and our program has set the Label text as we requested. Note that the program is still running. We can click as many times as we like with the same outcome. Only when we close the program by clicking on the top right **X** will it terminate.

Whilst we have only typed one line of code, Delphi has typed many for us. Let us first look at the main program code. Notice that we have added starting with the // comment identifier). These are ignored by the Delphi compiler, but help the coder understand the code.:

unit Unit1;

interface

uses
Windows, Messages, SysUtils, Variants, Classes, Graphics, Controls, Forms,
Dialogs, StdCtrls;

type
Tform1 = class(Tform)
Label: Tlabel; // The label we have added
Button1: Tbutton; // The button we have added
procedure Button1Click(Sender: Tobject);
private
{ **private** declarations }
public
{ **public** declarations }
end;

var

Form1: Tform1;

implementation

{$R *.dfm}

// The button action we have added
procedure Tform1.Button1Click(Sender: Tobject);
begin
Label1.Caption := 'Hello World'; // Label changed when button pressed
end;

end.

This code is called a **Unit** and is a Delphi module – one chunk of code. If you save this code, it will save in a file called **Unit1.pas** – a **Pascal** file. The unit comprises two main parts – the **interface** section, which tells what the unit does. And an **implementation** section that holds the code that implements the interface. Click on the **unit** keyword in the code to learn more.

Installing a virtual component library.

Go to http://www.uniworks.com/download.htm and download the evaluation copy(14 MB file).

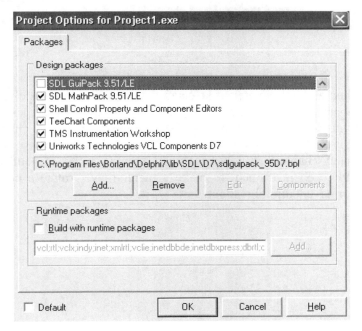

Click on the icon and the software components will automatically install in the Delphi environment. Open Delphi and click on 'components', 'Install packages' and the ticked box 'Uniworks technologies VCL Components D7' will be shown. Now click on the menu bar on tab 'Uniworks' and the virtual components will be presented on the bottom bar of the Delphi menu:

It is then a simple matter of selected the component with a click of the mouse and clicking again on the form in the chosen position. These instruments can then be scaled and their appearance changed by right clicking on the component and choosing properties. In the following chapters it will be shown how to write the software 'behind' each instrument to make a meter move, rotate a control knob or have a digital readout.

1.2 RS232 basics

RS232 is a serial connection technology standard adopted by the EIA (Electronic Industries Association) in 1960 and developed into the RS232C revision in 1969. Today it is still widely used for connecting PC's, terminals (VDU's), medical equipment, electronics test equipment, bar code readers and a multitude of monitoring and control equipment and of course PIC's.

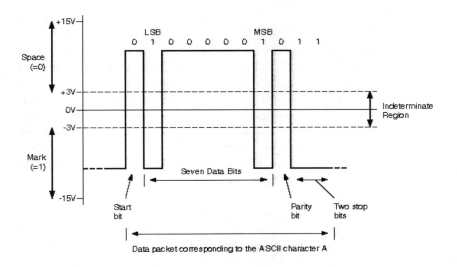

Typical representation of a RS232 signal

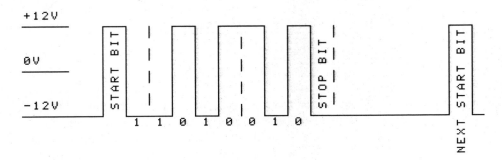

diagram showing the letter 'A' 0x41, 8 data bits, 1 start and stop bit

RS232 specifications, introduction

Communication as defined in the RS232 standard is an asynchronous serial communication method. The word serial means, that the information is sent one bit at a time. Asynchronous tells us that the information is not sent in predefined time slots. Data transfer can start at any given time and it is the task of the receiver to detect when a message starts and ends.

RS232 bit streams

The RS232 standard describes a communication method where information is sent bit by bit on a physical channel. The information must be broken up in data words. The length of a data word is variable. On PC's a length between 5 and 8 bits can be selected. This length is the netto information length of each word. For proper transfer additional bits are added for synchronisation and error checking purposes. It is important, that the transmitter and receiver use the same number of bits. Otherwise, the data word may be misinterpreted, or not recognized at all.

Data bits are sent with a predefined frequency, the baud rate. For a baud rate of 9600, the bit period is ~104 microseconds (1/9600 seconds). Both the transmitter and receiver must be programmed to use the same bit frequency. After the first bit is received, the receiver calculates at which moments the other data bits will be received. It will check the line voltage levels at those moments. With RS232, the line voltage level can have two states. The on state is also known as mark, the off state as space. No other line states are possible. When the line is idle, it is kept in the mark state.

Start bit

RS232 defines an asynchronous type of communication. This means, that sending of a data word can start on each moment. If starting at each moment is possible, this can pose some problems for the receiver to know which is the first bit to receive. To overcome this problem, each data word is started with an attention bit. This attention bit, also known as the start bit, is always identified by the space line level. Because the line is in mark state when idle, the start bit is easily recognized by the receiver.

Data bits

Directly following the start bit, the data bits are sent. A bit value 1 causes the line to go in mark state, the bit value 0 is represented by a space. The least significant bit is always the first bit sent.

Parity bit

For error detecting purposes, it is possible to add an extra bit to the data word automatically. The transmitter calculates the value of the bit depending on the information sent. The receiver performs the same calculation and checks if the actual parity bit value corresponds to the calculated value. The Parity bit is unused in all the examples.

Stop bits

Suppose that the receiver has missed the start bit because of noise on the transmission line. It started on the first following data bit with a space value. This causes garbled date to reach the receiver. A mechanism must be present to resynchronize the communication. To do this, framing is introduced. Framing means, that all the data bits and parity bit are contained in a frame of start and stop bits. The period of time lying between the start and stop bits is a constant defined by the baud rate and number of data and parity bits. The start bit has always space value, the stop bit always mark value. If the receiver detects a value other than mark when the stop bit should be present on the line, it knows that there is a synchronization failure. This causes a framing error condition in the receiving UART. The device then tries to resynchronize on new incoming bits.

For resynchronizing, the receiver scans the incoming data for valid start and stop bit pairs. This works, as long as there is enough variation in the bit patterns of the data words. The stop bit identifying the end of a data frame can have different lengths. Actually, it is not a real bit but a minimum period of time the line must be idle (mark state) at the end of each word. On PC's this period can have three lengths: the time equal to 1, 1.5 or 2 bits. 1.5 bits is only used with data words of 5 bits length and 2 only for longer words. A stop bit length of 1 bit is possible for all data word sizes.

RS232 physical properties

The RS232 standard describes a communication method capable of communicating in different environments. This has had its impact on the maximum allowable voltages etc. on the pins. In the original definition, the technical possibilities of that time were taken into account. The maximum baud rate was originally defined as 20 kbps. Today baud rates of 300 to 115200 are standard.

Voltages

The signal level of the RS232 pins can have two states. A high bit, or mark state is identified by a negative voltage and a low bit or space state uses a positive value. This might be a bit confusing, because in normal circumstances, high logical values are defined by high voltages also. The voltage limits are shown below.

RS232 voltage values

Space state (0)
+5 ... +15
+3 ... +25
Mark state (1)
-5 ... -15
-3 ... -25
Undefined
-3 ... +3

The maximum voltage swing the computer can generate on its port can have influence on the maximum cable length and communication speed that is allowed. Also, if the voltage difference is small, data distortion will occur sooner. For example, my Toshiba laptop mark's voltage is –9.3 V, compared to –11.5 V on my desktop computer. The laptop has difficulties to communicate with Mitsubishi PLC's in industrial environments with high noise levels where the desktop computer has no data errors at all using the same cable. Thus, even far beyond the minimum voltage levels, 2 volts extra can make a huge difference in communication quality. In the case examples in this book, most are powered by 5 volts or as little as 3.6 volts for a Lithium ion battery. This is within the RS232 specification of +3 to +15 volts. The FTDI and 4D USB/RS232 converters do not generate any –ve voltages on the RXD and TXD pins and therefore the RS232 voltage logic levels are +5v (logic 0) and 0v (logic 1) as shown below.

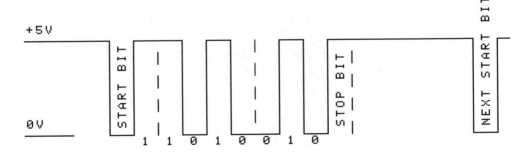

TTL level 'RS232' inverted signal for PIC's

RS232 is a definition for serial communication on a 1:1 base. RS232 defines the interface layer, but not the application layer. To use RS232 in a specific situation, application specific software must be written on devices on both ends of the connecting USB/RS232 cable. The developer is free to define the protocol used to communicate. RS232 ports can be either accessed directly by an application, or via a device driver in the operating system. This means that RS232 communication software must be written for the connecting PIC and also the PC 'front-end'. In this book, RS232 enabled software is used in the Delphi environment by using FTDI's RS232 functions or 'Portcontroller' RS232 procedures.

Pins 2 (RXD), 3(TXD) and pin 5(SIGNAL GROUND) are the only pins required for RS232 communications. The other pins known as hardware handshake lines were originally used for connecting to a modem and are now rarely used. However, when tied together, they provide a voltage of ~ 9 volts @ 20 mA which can be regulated by a 5V1 zener or 5 volt voltage regulator to supply a PIC.

Identical pin outs are found on the PROLIFIC/BELKIN USB RS232 converters and FTDI's EVAL232R so it can be seen that these convertors convert USB signals to RS232.

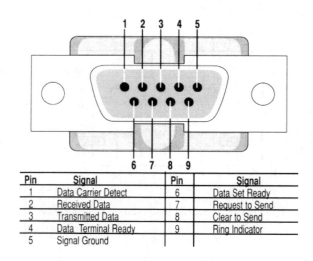

Pin	Signal	Pin	Signal
1	Data Carrier Detect	6	Data Set Ready
2	Received Data	7	Request to Send
3	Transmitted Data	8	Clear to Send
4	Data Terminal Ready	9	Ring Indicator
5	Signal Ground		

The RS232 'COM' port on a PC (9 pin D type male)

Maximum cable lengths

Cable length is one of the most discussed items in RS232 world. The standard has a clear answer, the maximum cable length is 50 feet, or the cable length equal to a capacitance of 2500 pF. The latter rule is often forgotten. This means that using a cable with low capacitance allows you to span longer distances without going beyond the limitations of the standard. If for example UTP CAT-5 cable is used with a typical capacitance of 17 pF/ft, the maximum allowed cable length is 147 feet. The cable length mentioned in the standard allows maximum communication speed to occur. If speed is reduced by a factor 2 or 4, the maximum length increases dramatically. Texas Instruments did some practical experiments years ago at different baud rates in order to test the maximum allowed cable lengths. Keep in mind, that the RS232 standard was originally developed for 20 kbps. But by halving the maximum communication speed, the allowed cable length increases a factor of ten!

RS232 cable length according to Texas Instruments:

Maximum cable length (feet)	Baud rate
50	19200
1000	9600
3000	2400

1.3 USB basics

Universal Serial Bus is a high speed connectivity standard enabling simple plug and play connections to devices such as modems, digital cameras, camcorders, keyboards and mice. USB was introduced in 1995 after being developed by major corporations including Intel, DEC, Microsoft, IBM and Compaq. The standard continues to be supported by many leading suppliers of computers and peripherals.

Topology:
The USB specification requires a host system to be equipped with an A type jack. The B type jack is typically found on peripheral devices requiring detachable cables. Therefore, detachable USB cables are configured as an A to B male combination which prevents improper bus configurations and topology miswiring.

Connectors

Type A Plug (4 position) | Type A Jack (4 position) | Type B Plug (4 position) | Type B Jack (4 position)

Mini Type B Plug (4 position) | Mini Type B Jack (4 position) | Mini Type B Plug (5 position) | Mini Type B Jack (5 position)

The aim of the USB-IF was to find a solution to the mixture of connection methods to the PC, in use at the time. We had serial ports, parallel ports, keyboard and mouse connections, joystick ports, midi ports and so on. And none of these satisfied the basic requirements of plug-and-play. Additionally many of these ports made use of a limited pool of PC resources, such as Hardware Interrupts, and DMA channels.

So the USB was developed as a new means to connect a large number of devices to the PC, and eventually to replace the 'legacy' serial and parallel ports. It was designed not to require specific Interrupt or DMA resources, and also to be 'hot-pluggable'. It was important that no special user-knowledge would be required to install a new device, and all devices would be distinguishable from all other devices, such that the correct driver software was always automatically used.

Architecture

The USB is based on a so-called 'tiered star topology' in which there is a single host controller and up to 127 'slave' devices. The host controller is connected to a hub, integrated within the PC, which allows a number of attachment points (often loosely referred to as ports). A further hub may be plugged into each of these attachment points, and so on. However there are limitations on this expansion.

As stated above a maximum of 127 devices (including hubs) may be connected. This is because the address field in a packet is 7 bits long, and the address 0 cannot be used as it has special significance. (In most systems the bus would be running out of bandwidth, or other resources, long before the 127 devices was reached.) A device can be plugged into a hub, and that hub can be plugged into another hub and so on. However the maximum number of tiers permitted is six.

The length of any cable is limited to 5 metres. This limitation is expressed in the specification in terms of cable delays etc, but 5 metres can be taken as the practical consequence of the specification. This means that a device cannot be further than 30 metres from the PC, and even to achieve that will involve 5 external hubs, of which at least 2 will need to be self-powered, so the USB is intended as a bus for devices near to the PC. For applications requiring distance from the PC, another form of connection is needed, such as Ethernet.

A 4-way USB hub

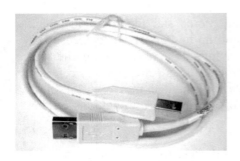

USB cable USB/RS232 converter

The advantage of using a USB/RS232 converter is that whereas software programming for RS232 is a relatively simple process, programming for USB interfacing is a non-trivial task and requires considerable expertise.In the case studies in this book, we can consider the USB interface as a 'black box' and need not concern ourselves with it's details.

1.4   USB RS232 serial and parallel bridges

There are basically two types of USB/RS232 adapters – one has internal logc inverters and voltage level converters such as the MAX232, to present a industry standard RS232 signal and function exclusively on a VCP (COM port). Typical of these are the BELKIN/PROLIFIC (PL2303) and FTDI's EVALRS232R (DLL and VCP) converters which are available at most computer suppliers. The other types operate on standard 5 volt logic signals and have no logic inverters. FTDI have a range of USB/RS232 converters such as the UB232R and the MM232R (DLL and VCP). 4D systems have an equivalent converter – the uUSB-MB which however will function only on a VCP port. The PROLIFIC etc type are only required when communicating over several hundred yards since signal attenuation increases with distance. However if the USB/RS232 interface circuit is within a few feet of the PC, simple TTL logic will work even at high baud rates.

BELKIN/PROLIFIC

FTDI MM232R   FTDI UB232R

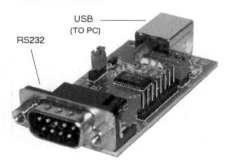

4D SYSTEMS uUSB   FTDI EVAL232R

USB-RS232 CONVERTERS

Also FTDI has a parallel USB/RS232 converter (The DLP-USB245M) which provides a parallel input/output bus for interfacing to 8 port lines of a PIC.

FTDI MM232R Schematic (courtesy of FTDI)

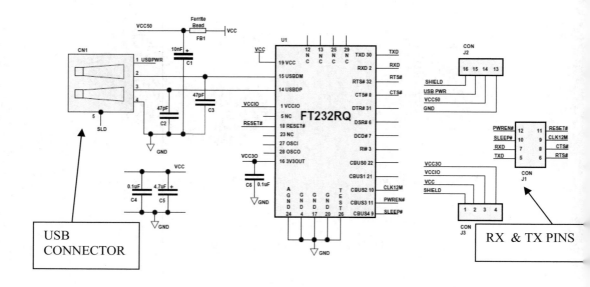

Based on FTDI's FT232RQ ASIC, this single chip converter contains a clock and an EEPROM.

1.2 Driver Support

Royalty-Free VIRTUAL COM PORT (VCP) DRIVERS for...	Royalty-Free D2XX *Direct* Drivers (USB Drivers + DLL S/W Interface)
• Windows 98, 98SE, ME, 2000, Server 2003, XP.	• Windows 98, 98SE, ME, 2000, Server 2003, XP.
• Windows Vista / Longhorn*	• Windows Vista / Longhorn*
• Windows XP 64-bit.*	• Windows XP 64-bit.*
• Windows XP Embedded.	• Windows XP Embedded.
• Windows CE.NET 4.2 & 5.0	• Windows CE.NET 4.2 & 5.0
• MAC OS 8 / 9, OS-X	• Linux 2.4 and greater
• Linux 2.4 and greater	

The drivers listed above are all available to download for free from the FTDI website. Various 3rd Party Drivers are also available for various other operating systems - see the FTDI website for details.

* Currently Under Development. Contact FTDI for availability.

1.3 Typical Applications

- USB to RS232 / RS422 / RS485 Converters
- Upgrading Legacy Peripherals to USB
- Cellular and Cordless Phone USB data transfer cables and interfaces
- Interfacing MCU / PLD / FPGA based designs to USB
- USB Audio and Low Bandwidth Video data transfer
- PDA to USB data transfer
- USB Smart Card Readers
- USB Instrumentation

- USB Industrial Control
- USB MP3 Player Interface
- USB FLASH Card Reader / Writers
- Set Top Box PC - USB interface
- USB Digital Camera Interface
- USB Hardware Modems
- USB Wireless Modems
- USB Bar Code Readers
- USB Software / Hardware Encryption Dongles

MM232R Mini USB-Serial UART Development Module Datasheet Version 1.00   © Future Technology Devices International Ltd. 2005

1.1 Hardware Features

- Single chip USB to asynchronous serial data transfer interface.
- Entire USB protocol handled on the chip - No USB-specific firmware programming required.
- UART interface support for 7 or 8 data bits, 1 or 2 stop bits and odd / even / mark / space / no parity.
- Fully assisted hardware or X-On / X-Off software handshaking.
- Data transfer rates from 300 baud to 3 Megabaud (RS422 / RS485 and at TTL levels) and 300 baud to 1 Megabaud (RS232).
- FTDI's royalty-free VCP and D2XX drivers eliminate the requirement for USB driver development in most cases.
- In-built support for event characters and line break condition.
- New USB FTDIChip-ID™ feature.
- New configurable CBUS I/O pins.
- Auto transmit buffer control for RS485 applications.
- Transmit and receive LED drive signals.
- New 48MHz, 24MHz, 12MHz, and 6MHz clock output signal Options for driving external MCU or FPGA.
- FIFO receive and transmit buffers for high data throughput.
- Adjustable receive buffer timeout.
- Synchronous and asynchronous bit bang mode interface options with RD# and WR# strobes.
- New CBUS bit bang mode option.
- Integrated 1024 bit internal EEPROM for storing USB VID, PID, serial number and product description strings, and CBUS I/O configuration.
- Device supplied preprogrammed with unique USB serial number.
- Support for USB suspend and resume.
- Support for bus powered, self powered, and high-power bus powered USB configurations.
- On board jumper allows for selection of USB bus powered supply or self powered supply.
- Integrated 3.3V level converter for USB I/O .
- Integrated level converter on UART and CBUS for interfacing to 5V - 1.8V Logic.
- On board jumper allows for selection of UART and CBUS interface IO voltage.
- True 5V / 3.3V / 2.8V / 1.8V CMOS drive output and TTL input.
- High I/O pin output drive option.
- Integrated USB resistors.
- Integrated power-on-reset circuit.
- Fully integrated clock - no external crystal, oscillator, or resonator required.
- Fully integrated AVCC supply filtering - No separate AVCC pin and no external R-C filter required.
- UART signal inversion option.
- USB bulk transfer mode.
- 3.3V to 5.25V Single Supply Operation.
- Low operating and USB suspend current.
- Low USB bandwidth consumption.
- UHCI / OHCI / EHCI host controller compatible
- USB 2.0 Full Speed compatible.
- -40°C to 85°C extended operating temperature range.
- Supplied on minature 18.0mm x 21.5mm PCB with 16 board header pins (on 2.25mm / 0.1" pitch).
- Connect to a PC via a standard USB A to B USB cable.
- Two 1x4 and one 2x4 PCB header sockets are supplied with the MM232R module, as standard.

1.3 Typical Applications

- USB to RS232 / RS422 / RS485 Converters
- Upgrading Legacy Peripherals to USB
- Cellular and Cordless Phone USB data transfer cables and interfaces
- Interfacing MCU / PLD / FPGA based designs to USB
- USB Audio and Low Bandwidth Video data transfer
- PDA to USB data transfer
- USB Smart Card Readers
- USB Instrumentation
- USB Industrial Control
- USB MP3 Player Interface
- USB FLASH Card Reader / Writers
- Set Top Box PC - USB interface
- USB Digital Camera Interface
- USB Hardware Modems
- USB Wireless Modems
- USB Bar Code Readers
- USB Software / Hardware Encryption Dongles

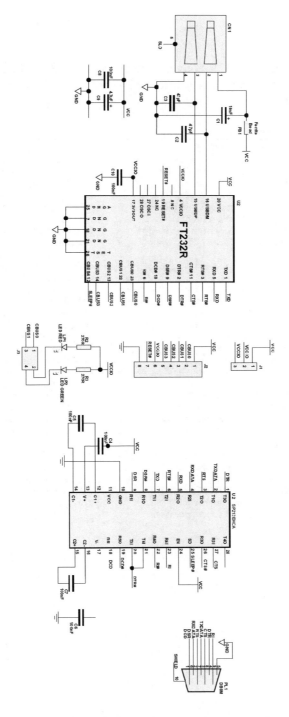

FTDI EVAL232R Schematic (courtesy of FTDI)

The FTDI EVAL232R converter contains a RS232 voltage level and logic inverter providing a true RS232 interface on a 9 pin D-type connector.

1.1 Hardware Features

- FT232RL provides single chip USB to asynchronous serial data transfer interface.
- Module provides USB to RS232 interface, with additional I/O pinout.
- Entire USB protocol handled on the FT232RL chip - No USB-specific firmware programming required.
- UART interface support for 7 or 8 data bits, 1 or 2 stop bits and odd / even / mark / space / no parity.
- Fully assisted hardware or X-On / X-Off software handshaking.
- Data transfer rates from 300 baud to 1 Megabaud at RS232 levels.
- FTDI's royalty-free VCP and D2XX drivers eliminate the requirement for USB driver development in most cases.
- In-built support for event characters and line break condition.
- New USB FTDIChip-ID™ feature.
- New configurable CBUS I/O pins.
- Transmit and receive LED's to indicate serial data transfer.
- New 48MHz, 24MHz, 12MHz, and 6MHz clock output signal Options.
- FIFO receive and transmit buffers for high data throughput.
- Adjustable receive buffer timeout.
- Synchronous and asynchronous bit bang mode interface options with RD# and WR# strobes.
- New CBUS bit bang mode option.
- Integrated 1024 bit internal EEPROM for storing USB VID, PID, serial number and product description strings, and CBUS I/O configuration.
- Device supplied preprogrammed with unique USB serial number.
- Support for USB suspend and resume.
- Module supply voltage comes from VCC - no external supply required.
- 5V USB power available to supply external devices.
- Integrated 3.3V level converter for USB I/O.
- Integrated level converter on UART and CBUS for interfacing to 5V - 1.8V Logic.
- On board jumper allows for selection of UART and CBUS interface IO voltage.
- True 5V / 3.3V / 2.8V / 1.8V CMOS drive output and TTL input.
- High I/O pin output drive option.
- Integrated USB resistors.
- Integrated power-on-reset circuit.
- Fully integrated clock - no external crystal, oscillator, or resonator required.
- Fully integrated AVCC supply filtering - No separate AVCC pin and no external R-C filter required.
- USB bulk transfer mode.
- Low operating and USB suspend current.
- Low USB bandwidth consumption.
- UHCI / OHCI / EHCI host controller compatible
- USB 2.0 Full Speed compatible.
- -40°C to 85°C extended operating temperature range.
- Supplied on minature 61.0mm x 32.0mm (2.40" x 1.26") PCB with 15 board header pins.
- Connect to a PC via a standard USB A to B USB cable.

1.5 PIC's and their ability to communicate using onboard UART and software 'bit-bang'

Many of the PIC's featured in this Book have on board UART (universal asynchronous receive transmit) hardware which enables the PIC to receive and transmit serial RS232 data under software control. The UART is a simple parallel/serial and serial/parallel hardware converter inside the PIC whereby the parallel 8 bit PIC transmit data is transmitted serially from a port pin and conversely the received serial RS232 data is received on a single port pin and converted to a parallel 8 bit data byte inside the PIC. All the following example programs for various PIC's can be downloaded from the ELEKTOR.COM website and can be located in Folder 'USART'

In the case of the PIC10F,PIC12F and some of the PIC16F Family, a UART is is not included and a software 'big bang' method is used to replicate the hardware UART. In this case, the 8 bit data to be transmitted is placed in a buffer register and rotated bit by bit into the carry flag. The bit on the transmit pin of the PIC is then set to a logic 1 or cleared to 0 dependent on wether the carry flag is set or cleared. All 8 bits are rotated through the buffer with a precise delay between each bit dependent on the bit rate. In addition a start bit is issued at the start of the process and a stop bit is added once all 8 bits have been transmitted.

For a serial byte to be received, the PIC pin used for reception is initially polled for the beginning of a start bit, after which a time period of 1.5 bits is introduced so that the logic level of the succesive 8 bits is polled in the middle of the bit period. A time period of 1 bit period is allocated between each bit received. As the 8 bits are received one by one, bit 7 of a buffer is cleared or set dependent on the logic level of the received bit and the buffer contents are rotated by 1, gradually building a complete 8 bit data byte in the buffer.

Since we are communicating with a PC with a USB port,this serial RS232 data needs to be converted by a USB RS232 converter. As seen previously, the FTDI UBRS232 and 4D uUSB converters do not have inbuilt logic conversion and for PIC's with on board UART's , the converter RXD and TXD pins are simply connected to the RC6/TX and RC7/RX of the PIC. The logic levels of the received/transmitted data are not RS232 voltage levels and operate with standard 5 volt (or lower) TTL logic. For PIC's which use the 'bit bang' method of software generated RS232 data, the PROLIFIC/BELKIN USB RS232 and FTDI's EVALRS232R converters have built in logic and standard RS232 voltage level conversion,so that care is needed in producing the correct 'inverted logic' in the software. Of course in this case, any PIC pin may be selected to receive and transmit data. The following examples shown below are of various PIC's and their software to achieve bidirectional communication with a PC. A terminal emulator is used to test each circuit. Pressing say the 'A' key on the PC keyboard will send the character 'A' to the PIC where it is received and re-tranmitted to the PC to demonstrate that the communication link is valid.

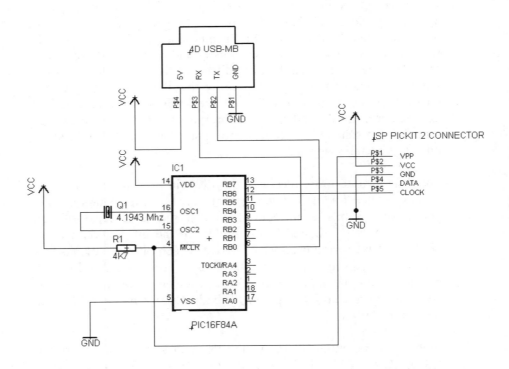

PIC16F84A serial interface

A PIC16F84A and not the PIC16F84 must be used since the later is not supported by the PICKIT 2. The 4D converter can be replaced by the FTDI UB232R.

;pic16f84-txd-rxd-uUSB..ASM

;9600 BAUD 4.1943 Mhz   4D uUSB comvertor

```
        LIST    p=16F84 ; PIC16F84 is the target processor

TXD     EQU 3       ;RB3 TRANSMIT
RXD     EQU 0       ;RB0 RECEIVE
PC      EQU 2
STATUS EQU 3        ;STATUS REGISTER

PORT_B EQU 6        ;PORT B

DLYCNT EQU 010H
BUFFER EQU 011H
COUNTR EQU 012H
INVERT EQU 015H
COUNTER EQU 016H
DATABYTE EQU 017H


BAUD_1 EQU .31    ; 9600 BAUD 1 BIT PERIOD
BAUD_1ST EQU .46   ;1.5 BIT PERIOD

    ORG 0
    CALL DELAY
    NOP
    MOVLW B'00000001'  ;RB3 TXD, RB0 RXD
    TRIS PORT_B

    MOVLW B'00000111'
    OPTION

    BCF PORT_B,TXD      ;STOP BIT LOW
CYCLE

    CALL RECEIVE    ;GET BYTE
    MOVF DATABYTE,0

;  MOVLW 'T'

    CALL TRANSMIT
    GOTO CYCLE


TRANSMIT

    MOVWF BUFFER      ;IN FILE REGISTER A
```

```
         MOVLW 8        ;8 DATA BITS
         MOVWF COUNTR


         BCF PORT_B,TXD    ;SEND START BIT TXD LINE HIGH
NEXT  CALL DELAY        ;104 uS DELAY
         RRF BUFFER       ;ROTATE BUFFER
         BTFSS STATUS,0   ;TEST CARRY FLAG
         BCF PORT_B,TXD
         BTFSC STATUS,0   ;TEST AND TRANSMIT BIT
         BSF PORT_B,TXD
         DECFSZ COUNTR    ;UNTIL ALL 8 BITS TRANSMITTED
         GOTO NEXT
         CALL DELAY
         BSF PORT_B,TXD   ;STOP BIT LOW
         CALL DELAY

         RETLW 0


RECEIVE

START_BIT
   CLRWDT
   BTFSC PORT_B,RXD
   GOTO START_BIT

RXD_DATA
     MOVLW 8       ;8 DATA BITS
     MOVWF COUNTR
     CALL DELAY1ST    ;1.5 BITS DELAY
     CLRF BUFFER      ;CLEAR BUFFER

NEXT_BITR
   BSF STATUS,0
   RRF BUFFER
   BTFSS PORT_B,RXD
   BCF BUFFER,7
   CALL DELAY
   DECFSZ COUNTR,1
   GOTO NEXT_BITR

 ; MOVLW 0FFH
 ; XORWF BUFFER,0
 ; MOVWF DATABYTE      ;SAVE RECEIVED DATA
   MOVF BUFFER,0
   MOVWF DATABYTE

   RETLW 0
 DELAY              ;104 uS DELAY
     MOVLW BAUD_1
     MOVWF DLYCNT
REDO DECFSZ DLYCNT,1
      goto REDO
      CLRWDT
      retlw 0
```

```
DELAY1ST
    MOVLW BAUD_1ST
    MOVWF DLYCNT
REDO1ST DECFSZ DLYCNT,1
    goto REDO1ST
    CLRWDT
    retlw 0

    END
```

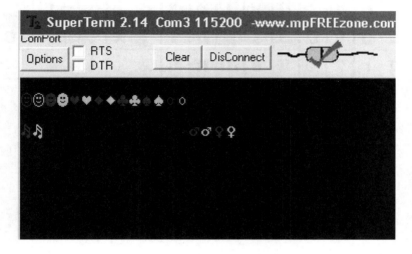

Ctrl A,Ctrls B,Ctrl C etc will output 0x1,0x2,0x3 etc. Note These are not printable standard ASCII characters

A terminal emulator showing echo'ed characters from the PIC

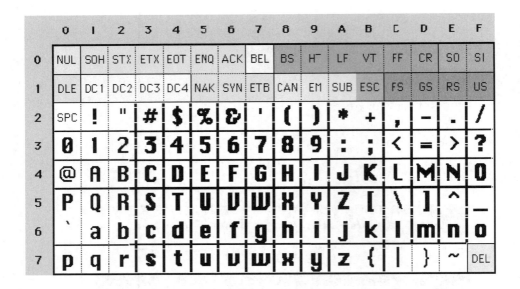

The standard ASCII character set 0x00 to 0x7F

The table above shows how each alpha numeric character on the keyboard in encoded into a byte. For example, when the 'A' key is pressed, 0x41 is sent to the serial port as a RS232 byte. For the letter a', 0x61 is sent.

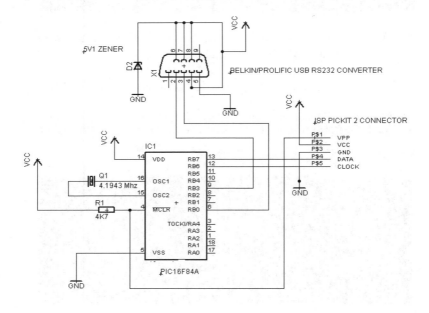

PIC16F84A serial interface with BELKIN USB RS232 adapter

Note that the unused hardware handshake lines are tied together on the RS232 connector and clamped to 5 volts by a zener to provide the VCC for the circuit.

```
;pic16f84-txd-rxd.ASM


;9600 BAUD 4.1943 Mhz belkin/prolific usb comvertEr

      LIST   p=16F84A ; PIC16F84A is the target processor

TXD    EQU 3      ;RB3 TRANSMIT
RXD    EQU 0      ;RB0 RECEIVE
PC     EQU 2
STATUS EQU 3      ;STATUS REGISTER

PORT_B EQU 6      ;PORT B

DLYCNT EQU 010H
BUFFER EQU 011H
COUNTR EQU 012H
INVERT EQU 015H
COUNTER EQU 016H
DATABYTE EQU 017H



BAUD_1 EQU .31    ; 9600 BAUD 1 BIT PERIOD
BAUD_1^ST EQU .46   ;1.5 BIT PERIOD

    ORG 0
   CALL DELAY
   NOP
   MOVLW B'00000001'  ;RB3 TXD, RB0 RXD
   TRIS PORT_B

   MOVLW B'00000111'
   OPTION

  BCF PORT_B,TXD     ;STOP BIT LOW

CYCLE


  CALL RECEIVE   ;GET BYTE
  MOVF DATABYTE,0

; MOVLW 'T'
  CALL CONVERT
  CALL TRANSMIT
  GOTO CYCLE
```

```
CONVERT
   XORLW 0FFH
   MOVWF BUFFER    ;IN FILE REGISTER A
   MOVLW 8         ;8 DATA BITS
   MOVWF COUNTR
   RETLW 0

TRANSMIT

   BSF PORT_B,TXD   ;SEND START BIT TXD LINE HIGH
NEXT  CALL DELAY    ;104 uS DELAY
   RRF BUFFER       ;ROTATE BUFFER
   BTFSC STATUS,0   ;TEST CARRY FLAG
   BSF PORT_B,TXD
   BTFSS STATUS,0   ;TEST AND TRANSMIT BIT
   BCF PORT_B,TXD
   DECFSZ COUNTR    ;UNTIL ALL 8 BITS TRANSMITTED
   GOTO NEXT
   CALL DELAY
   BCF PORT_B,TXD   ;STOP BIT LOW
   CALL DELAY

   RETLW 0


RECEIVE

START_BIT
   CLRWDT
   BTFSS PORT_B,RXD
   GOTO START_BIT

RXD_DATA
   MOVLW 8          ;8 DATA BITS
   MOVWF COUNTR
   CALL DELAY1ST    ;1.5 BITS DELAY
   CLRF BUFFER      ;CLEAR BUFFER

NEXT_BITR
   BCF STATUS,0
   RRF BUFFER
   BTFSC PORT_B,RXD
   BSF BUFFER,7
   CALL DELAY
   DECFSZ COUNTR,1
   GOTO NEXT_BITR

   MOVLW 0FFH
   XORWF BUFFER,0
   MOVWF DATABYTE       ;SAVE RECEIVED DATA
   RETLW 0
```

```
DELAY              ;104 uS DELAY
    MOVLW BAUD_1
    MOVWF DLYCNT
REDO DECFSZ DLYCNT,1
    goto REDO
    CLRWDT
    retlw 0


DELAY1ST
    MOVLW BAUD_1ST
    MOVWF DLYCNT
REDO1ST DECFSZ DLYCNT,1
    goto REDO1ST
    CLRWDT
    retlw 0



    END
```

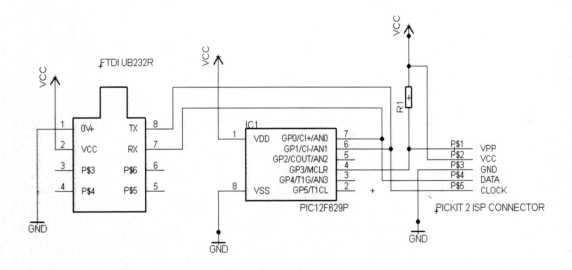

PIC12F629 serial interface

No xtal is required since the PIC uses it's own internal 4 Mhz RC oscillator. Note that this fuse configuration is only suitable for low i.e < 4800 baud rates, since due to temperature changes, the bit period can vary as much as 10% and thus communications will be unreliable. The FTDI converter can be replaced with 4D's converter which is electrically identical.

```
;pic12f629-txd-rxd-uUSB.ASM

;internal rc clock 2400 baud,   4D uUSB converter or FTDI uUB232R converter

        list    p=12F629    ; list directive to define processor
        #include <p12f629.inc>   ; processor specific variable definitions

        __CONFIG _CP_OFF & _WDT_OFF & _BODEN_ON & _PWRTE_OFF
& _INTRC_OSC_NOCLKOUT & _MCLRE_OFF & _CPD_OFF


RTCC    EQU 1      ;RTCC
PC      EQU 2
STATUS  EQU 3      ;STATUS REGISTER
BUFFER EQU 020H
COUNTR EQU 021H
DLYCNT EQU 023H
DATABYTE EQU 025H

TXD EQU 0
GPIO EQU 5
CMCON EQU 019H
ANSEL EQU 09FH


    movwf  OSCCAL


        BCF STATUS,RP0      ;BANK 0
        CLRF GPIO           ;INIT GPIO
        MOVLW 7
        MOVWF CMCON         ;DISABLE COMPARATOR
        BSF STATUS,RP0              ;BANK1
        CLRF ANSEL     ;DIGITAL IO

        MOVLW  2      ;GP0 TXD, GP1-RXD
        MOVWF  TRISIO

    MOVLW B'00000111';  PRESCALER 00000111 (/256)
    MOVWF  OPTION_REG

        BCF STATUS,RP0              ; change back to PORT memory bank

CYCLE
    CALL RECEIVE_DATA
    MOVF DATABYTE,W
;   MOVLW 'T'
    CALL TRANSMIT
```

```
        GOTO CYCLE



TRANSMIT
   MOVWF BUFFER

   MOVLW 8       ;8 DATA BITS
   MOVWF COUNTR

   BCF GPIO,TXD    ;SEND START BIT TXD LINE HIGH
NEXT
    CALL DELAY
             ;416 uS DELAY
   RRF BUFFER,1    ;ROTATE BUFFER
   BTFSS STATUS,0  ;TEST CARRY FLAG
   BCF GPIO,TXD
   BTFSC STATUS,0   ;TEST AND TRANSMIT BIT
   BSF GPIO,0
   DECFSZ COUNTR,1  ;UNTIL ALL 8 BITS TRANSMITTED
   GOTO NEXT
   CALL DELAY
   BSF GPIO,TXD   ;STOP BIT
   CALL DELAY
   RETURN



RECEIVE_DATA

START_BIT
   CLRWDT
   BTFSC GPIO,1
   GOTO START_BIT
   MOVLW 8       ;8 DATA BITS
   MOVWF COUNTR
   CALL DELAY1ST     ;1.5 BITS DELAY
   CLRF BUFFER     ;CLEAR BUFFER

NEXTR
   BSF STATUS,0
   RRF BUFFER,1
   BTFSS GPIO,1
   BCF BUFFER,7
    CALL DELAY
   DECFSZ COUNTR,1
   GOTO NEXTR

   MOVF BUFFER,0
   MOVWF DATABYTE

   RETURN
```

```
DELAY
    MOVLW .135
    MOVWF DLYCNT
REDO DECFSZ DLYCNT,1
    goto REDO
    CLRWDT
    NOP
    RETURN

DELAY1ST

    MOVLW .202
    MOVWF DLYCNT
REDO1ST DECFSZ DLYCNT,1
    goto REDO1ST
    CLRWDT
    RETURN

    END
```

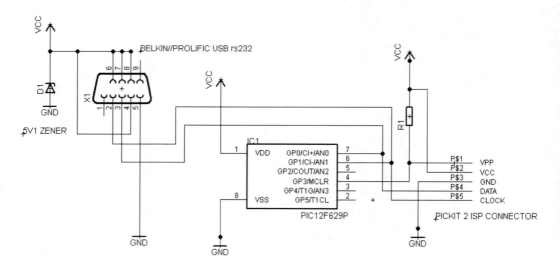

PIC12F629 serial interface with PROLIFIC USB RS232

;pic12f629-txd-rxd.ASM

;internal rc clock 2400 baud, prolific/BELKIN usb convertor

```
        list    p=12F629    ; list directive to define processor
        #include <p12f629.inc>   ; processor specific variable definitions

        __CONFIG _CP_OFF & _WDT_OFF & _BODEN_ON & _PWRTE_OFF
&_INTRC_OSC_NOCLKOUT & _MCLRE_OFF & _CPD_OFF


RTCC    EQU 1   ;RTCC
PC      EQU 2
STATUS  EQU 3   ;STATUS REGISTER
BUFFER EQU 020H
COUNTR EQU 021H
DLYCNT EQU 023H
DATABYTE EQU 025H

TXD  EQU 0
GPIO EQU 5
CMCON EQU 019H
ANSEL EQU 09FH


   movwf   OSCCAL


        BCF STATUS,RP0    ;BANK 0
        CLRF GPIO         ;INIT GPIO
        MOVLW 7
        MOVWF CMCON       ;DISABLE COMPARATOR
        BSF STATUS,RP0            ;BANK1
        CLRF ANSEL        ;DIGITAL IO

        MOVLW  2          ;GP0 TXD, GP1-RXD
        MOVWF  TRISIO

   MOVLW B'00000111';  PRESCALER 00000111 (/256)
   MOVWF OPTION_REG

        BCF STATUS,RP0             ; change back to PORT memory bank


CYCLE
   CALL RECEIVE_DATA
   MOVF DATABYTE,W

   ; MOVLW 'T'
   CALL CONVERT
   CALL TRANSMIT
   GOTO CYCLE




CONVERT
   XORLW 0FFH
   MOVWF BUFFER    ;IN FILE REGISTER A
   MOVLW 8         ;8 DATA BITS
```

```
        MOVWF COUNTR
        RETURN

TRANSMIT
   BSF GPIO,TXD      ;SEND START BIT TXD LINE HIGH
NEXT
    CALL DELAY
            ;416 uS DELAY
    RRF BUFFER,1    ;ROTATE BUFFER
    BTFSC STATUS,0   ;TEST CARRY FLAG
    BSF GPIO,TXD
    BTFSS STATUS,0   ;TEST AND TRANSMIT BIT
    BCF GPIO,0
    DECFSZ COUNTR,1   ;UNTIL ALL 8 BITS TRANSMITTED
    GOTO NEXT
    CALL DELAY
    BCF GPIO,TXD   ;STOP BIT
    CALL DELAY
    RETURN




       RECEIVE_DATA

START_BIT
   CLRWDT
   BTFSS GPIO,1
   GOTO START_BIT
   MOVLW 8        ;8 DATA BITS
   MOVWF COUNTR
   CALL DELAY1ST    ;1.5 BITS DELAY
   CLRF BUFFER      ;CLEAR BUFFER

NEXTR
   BCF STATUS,0
   RRF BUFFER,1
   BTFSC GPIO,1
   BSF BUFFER,7
    CALL DELAY
   DECFSZ COUNTR,1
   GOTO NEXTR

   MOVLW 0FFH
   XORWF BUFFER,0
   MOVWF DATABYTE


        RETURN

DELAY
     MOVLW .135
     MOVWF DLYCNT
REDO DECFSZ DLYCNT,1
    goto REDO
```

```
        CLRWDT
         NOP
        RETURN

    DELAY1ST

        MOVLW .202
        MOVWF DLYCNT
REDO1ST DECFSZ DLYCNT,1
        goto REDO1ST
        CLRWDT
        RETURN

    END
```

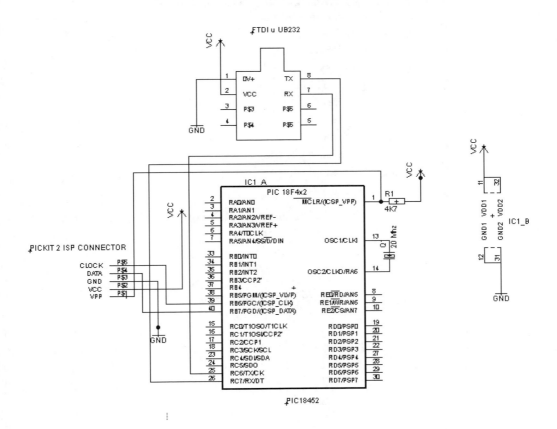

PIC18F452 serial interface

The RC6/TX and RC7/RX pins are common to most PIC16F and 18F PIC's.This time the program is writen in C and not assembler and has to be compiled in MPLAB'S C18 environment.Note that a 20 Mhz xtal is used to produce a baud rate of 115200.

Project 18f452_uart_HARD.mcw

//PROGRAM p18f452.c

```
// 115200 baud hard UART  20 Mhz clock, FDTI uUSB convertor
#include <p18f452.h>
#include <delays.h>
#include <usart.h>
extern void SleepMS(int MilliSeconds);

void delay10( char n);
 void main(void) ;
 void main(void)
  {
    int  i ;
    int data;
```

```
TRISC =0x80;// PORT RC6/TX OUTPUT, RC7/RX INPUT

OpenUSART(      USART_TX_INT_OFF & USART_RX_INT_OFF &
                        USART_ASYNCH_MODE                    &
    USART_EIGHT_BIT &
                    USART_CONT_RX & USART_BRGH_HIGH, 10
        );//115200 ,129 for 9600 baud



startx:
    while(!DataRdyUSART());
                        data = ReadUSART();
                        WriteUSART(data);
            while(BusyUSART());

    goto startx;


}
```

TABLE 16-5: BAUD RATES FOR ASYNCHRONOUS MODE (BRGH = 1)

BAUD RATE (Kbps)	Fosc = 40 MHz			33 MHz			25 MHz			20 MHz		
	KBAUD	% ERROR	SPBRG value (decimal)	KBAUD	% ERROR	SPBRG value (decimal)	KBAUD	% ERROR	SPBRG value (decimal)	KBAUD	% ERROR	SPBRG value (decimal)
0.3	NA	-	-	NA	-	-	NA	-	-	NA	-	-
1.2	NA	-	-	NA	-	-	NA	-	-	NA	-	-
2.4	NA	-	-	NA	-	-	NA	-	-	NA	-	-
9.6	NA	-	-	9.60	-0.07	214	9.59	-0.15	162	9.62	+0.16	129
19.2	19.23	+0.16	129	19.28	+0.39	106	19.30	+0.47	80	19.23	+0.16	64
76.8	75.76	-1.36	32	76.39	-0.54	26	78.13	+1.73	19	78.13	+1.73	15
96	96.15	+0.16	25	98.21	+2.31	20	97.66	+1.73	15	96.15	+0.16	12
300	312.50	+4.17	7	294.64	-1.79	6	312.50	+4.17	4	312.50	+4.17	3
500	500	0	4	515.63	+3.13	3	520.83	+4.17	2	416.67	-16.67	2
HIGH	2500	-	0	2062.50	-	0	1562.50	-	0	1250	-	0
LOW	9.77	-	255	8,06	-	255	6.10	-	255	4.88	-	255

BAUD RATE chart

The SPBRG value (BRGH_HIGH, 10) in the C code above specifies the baud rate. However this has to be calculated since the 115200 baud rate in not included in the above chart. The Microchip PIC18F452 data sheet expalins this in section 'UART'. For a baud rate of 9600 and with a clock of 20 Mhz, it can be seen that a SPBRG value of 129 is required.

PIC16F887- MICROCHIPS 44 PIN DEMO BOARD SERIAL INTERFACE

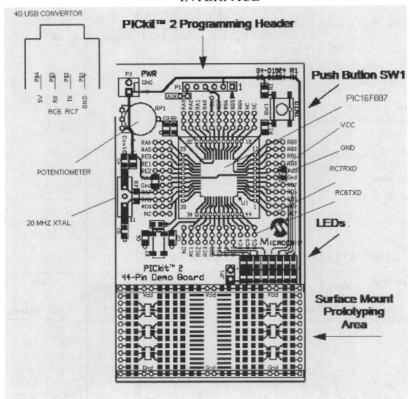

Here, the PIC16F887's UART is programmed in assembler. In addition to receiving and sending data, the board's 8 LED's also display the data in binary form. The 4D systems uUSB-MB connects to the RC6 and RC7 port lines of the board.

;pic16f887.ASM

;115200 baud, 20 Mhz clock 4D uUSB convertor

#include <p16F887.inc>
    __CONFIG   _CONFIG1, LVPOFF & FCMENOFF & IESOOFF & BOROFF & CPDOFF & CPOFF & MCLREOFF & PWRTEON & WDTOFF &   HSOSC
    __CONFIG   _CONFIG2, WRTOFF & _BOR21V

```
        BSF    STATUS, RP0  ; select bank 1
        movlw  b'10000000'  ; Setup port C for serial port.
                            ; TRISC<7>=1 and TRISC<6>=0.

        Movwf  TRISC

        clrf   TRISD        ; Make PortD all output
        clrf   TRISC

        BSF    TXSTA,BRGH   ; High BAUDrate
        MOVLW  .10  ; 115200 baud @ 20 Mhz Fosc +0.16% err


        MOVWF  SPBRG   ; load baudrate register

        BCF    TXSTA,SYNC   ; enable asynchronous transmission
        BSF    TXSTA,TXEN   ; enable transmission
        BCF    PIE1,RCIE    ; disable receive interrupt
        BCF    STATUS, RP0  ; back to bank 0
        BSF    RCSTA, SPEN  ; Enable serial port.
        BSF    RCSTA, CREN  ; Enable continuous reception

repeat:
    call receive

    goto repeat
receive:
wait:
  btfss PIR1,RCIF
  goto wait

    movf RCREG,W
    movwf PORTD ; DISPLAY THE DATA ON THE 8 LED'S

wait2:
    btfss PIR1,TXIF   ;transmit data
    goto wait2
    movwf TXREG
    retlw 0

    end
```

If we now install 'Portcontroller' described later in this book, and run executable application 'Numberxvcp'found in Folder 'PIC16F887' we can test the above PIC software from the Delphi program:

Finally we show how very low power consumption is achieved by using a PIC12C508, a 32.768 Khz watch crystal and a baud rate of 300.

```
PIC12C508-TXD.ASM

;300 baud ,32.768 Khz clock


    LIST   p=12C508



TXD    EQU 0     ;GPIO,0 TRANSMIT SERIAL RS232
RTCC   EQU 1
PC     EQU 2
STATUS EQU 3     ;STATUS REGISTER
GPIO EQU 6

DLYCNT  EQU 010H
BUFFER  EQU 011H ;TRANSMIT/RECEIVE BUFFER
COUNTR  EQU 012H ;TIMING COUNTER


    ORG 0
START

   CLRF GPIO
```

```
        MOVLW 0
        TRIS GPIO


        BCF GPIO,TXD    ;STOP BIT

CYCLE

        MOVLW 'T'
        CALL CONVERT
        CALL TRANSMIT




        GOTO CYCLE


CONVERT
        XORLW 0FFH
        MOVWF BUFFER
        MOVLW 8
        MOVWF COUNTR
        RETLW 0


TRANSMIT
        NOP
        BSF GPIO,TXD    ;SEND START BIT TXD LINE HIGH
NEXT CALL DELAY_3
        RRF BUFFER,1    ;ROTATE BUFFER
        BTFSC STATUS,0  ;TEST CARRY FLAG
        BSF GPIO,TXD
        BTFSS STATUS,0  ;TEST AND TRANSMIT BIT
        BCF GPIO,TXD
        DECFSZ COUNTR   ;UNTIL ALL 8 BITS TRANSMITTED
        GOTO NEXT

        CALL DELAY_3
        BCF GPIO,TXD    ;STOP BIT

        CALL DELAY_3

        RETLW 0


DELAY_3
        NOP
        NOP
        NOP
        NOP
        NOP
        NOP
        NOP
        NOP
        NOP
```

```
NOP
NOP
NOP
NOP
NOP
NOP
NOP
NOP
RETLW 0


END
```

1.6 Terminal emulation

A terminal emulator is software that runs on a PC to emulate a 'dumb' terminal, allowing RS232 ASCII text based characters to be received and displayed and also to be transmitted. Now that many PC's do not have an RS232 port, a USB to RS232 adaptor is used for RS232 communications to be achieved from the PC via the USB port to an external PIC. To use a terminal emulator, an active COM port number e.g. COM1 has to be selected in the settings as well as the baud rate. e.g. 9600 baud. The RS232 protocol also specifies the number of data bits, number of stop bits and even/odd parity. All of the examples in this book use 8 data bits, no parity and 1 stop bit. There are occasions when 2 stop bits are used when there is a large software overhead after a character is received, and the 2nd stop bit allows a 'breathing' space of 104 microseconds for a baud rate of 9600 which gives time for the software to execute.

There are many terminal emulations programs available including 'HyperTerminal' which comes installed with Windows XP but not with Vista. A favourite terminal of the authors is 'Superterm 2.14', although any terminal emulation program will provide the same functionality.

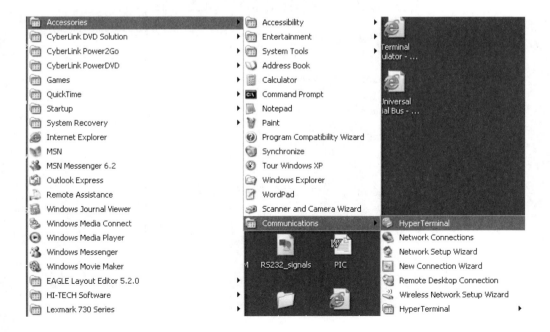

To invoke the PC's 'HyperTerminal' software click on 'Programs', 'Accessories', 'Communications', 'Hyperterminal'.

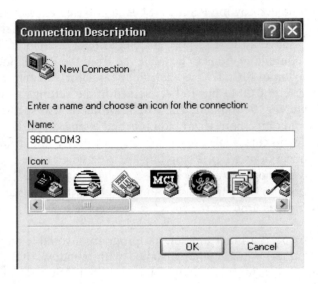

Enter a name for the new connection.

The COM port number is selected.

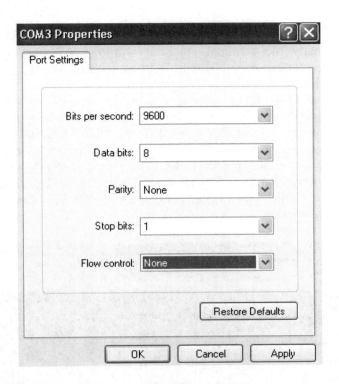

The RS232 communications protocol is selected.

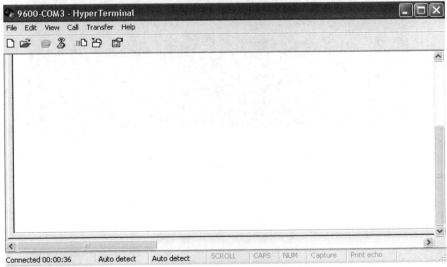

And finally the terminal emulator window is opened, allowing characters to be entered on the keyboard which will be sent and received from the PC.

The 'SuperTerm 2.14' emulator is easier to setup since, if an inactive COM port is selected, it gives the user the ability to change it immediatedly. Note, that when different USB/RS232 adapters are being used as virtual RS232 COM ports and are being continually plugged and unplugged from the USB port, the COM port active number may change and either the PC will have to be rebooted or the COM number can be changed in the PC's 'SYSTEM properties' hardware window.

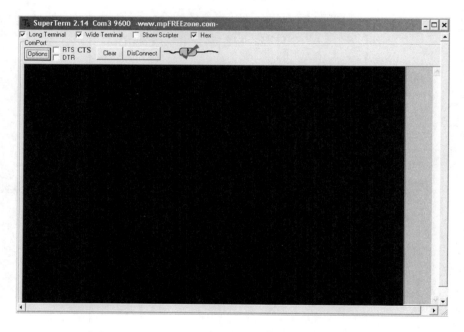

The Superterm Terminal emulation Window

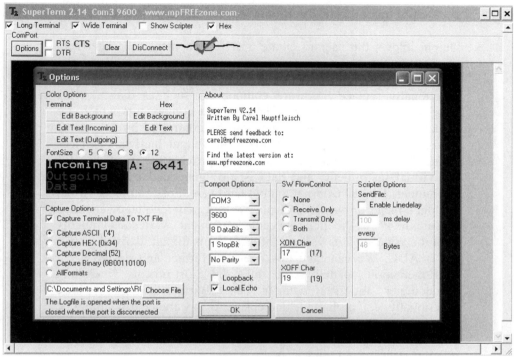

The 'Options' window

2 Delphi basics. Introduction to virtual instrumentation

Delphi allows the user to create a GUI (graphical user interface) application that runs in a Windows environment and allows the creation of virtual instrument computer programs that visually and functionally replace hardware components such as dials, knobs, gauges, meters and displays. These virtual components which are displayed on the computer screen can be controlled by use of the mouse or keyboard. External input and ouput hardware devices connected via the computer's USB port provide the input data for the virtual instruments and data is also generated by the virtual instruments for the external output hardware. For example a virtual analogue meter on the PC's display will display a voltage generated outside the PC as the meter's needle movement varies in sympathy with the voltage. This voltage information is digital, derived from a PIC's ADC. We can take this one step further and design a car instrumentation panel with all the instruments including speedometer and rev counter being virtual displays whose functionality e.g. mph range and cosmetic appearance can simply be changed in software. This means there are no moving parts to wear out and being digital it is simple to interface to external microprocessors and transducers.

All of the case studies in this Book utilise Microchips embedded microcontroller PIC's to get data to the PC's virtual components via the USB port and conversely to get data from the virtual components to external devices. This provides a comprehensive data and control system for a PC for data aquisition and control.

Delphi has limited virtual components and designing a component from scratch is a complex process. Therefore much use is made of 3rd party Delphi virtual component suppliers for the case studies in this Book. These components are easy to use and only require a single variable e.g. (Component.Value) consisting of a Byte (00-FFh) to control the simulation of the instrument. This chapter gives an introduction to producing a simulation of a virtual thermometer, a component to change the colour of a button, and with the aid of the Abacus virtual component library an example of a virtual flashing LED's.

2.1 A simple virtual thermometer

Our first virtual instrument is a simulation of a thermometer which displays the temperature digitally and as an analogue virtual instrument. A scrollbar controlled by the mouse generates a value of between 0 and 255 – this determines the hight of the red column i.e. the temperature.It can be seen that different scaling can be used (in this case divide by 7) and that the virtual thermometer can be any colour or size simply by editing the Delphi software. Later on, we shall see how to create a real world virtual thermometer which displays an accurate temperature derived from an external temperature transducer i.c.

The simple thermometer displayed on the screen.

Delphi Program simple_thermometerx.

Unit simple_thermometer;

interface

uses
 Windows, Messages, SysUtils, Classes, Graphics, Controls, Forms, Dialogs,
 OleCtrls, StdCtrls;

type

```
Tform1 = class(Tform)
  Label1: Tlabel;
  ScrollBar1: TscrollBar;

  procedure FormCreate(Sender: Tobject);
  procedure ScrollBar1Change(Sender: Tobject);
private
  { Private declarations }
public
  { Public declarations }
end;

var
  Form1: Tform1;

  text:byte;
  dat:byte;
  s:string;
implementation

{$R *.DFM}

procedure Tform1.FormCreate(Sender: Tobject);
begin
Form1.Canvas.Pen.Width:=10;
Form1.Canvas.Pen.color:=clsilver;
Form1.Canvas.Pen.color:=clRED;
end;

procedure Tform1.ScrollBar1Change(Sender: Tobject);
var
x:byte;
a:real;
Optval:byte;
begin
x:=Scrollbar1.position;
 Form1.Canvas.Pen.Width:=6;
Form1.Canvas.Pen.color:=clGREEN;
 Canvas.Moveto(50,300);
 Canvas.LineTo(50,45);
Form1.Canvas.Pen.color:=clRED;
Canvas.Moveto (50,300);
Canvas.LineTo(50,300-x);
a:=x;
a:=a/7;

Str(a:2:1,s);
Caption:=s;
Font.Color:=clWHITE;

Canvas.TextOut(100,200,'     ');
Canvas.TextOut(40,310,s);
end;

end.
```

2.2 Abacus virtual components

Our next virtual instrument example show 2 flashing LED's with software selectable LED colour, size and flash rate using the Abled component from the Abaecker Abakus professional set of Codegear Delphi real time components for virtual instrumentation. The library includes meters, bars (gauge), with linear or log(10) scaling,digital indicators (time and value), tank displays, trend graphical records, dial(knob) sliders, buttons, switches and LED indicators. Once installed in the Delphi environment, the Abacus AbBinary graphical element collection can be found which includes the Abled LED component.

The Abled Object inspector.

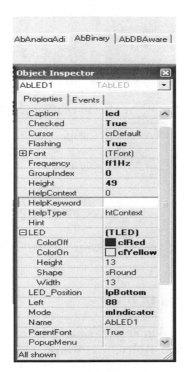

Some Abakus VCL components.

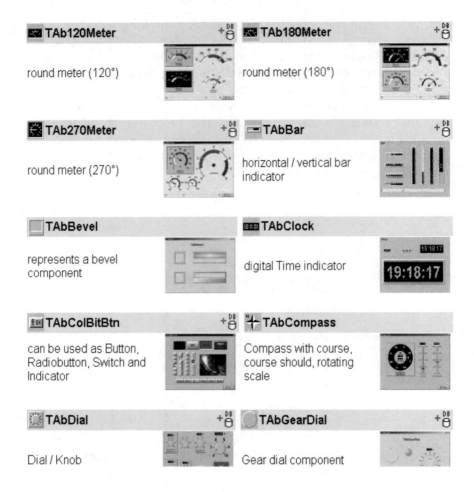

2.3 Virtual Flashing LED's

Program LED1x found in Folder 'simple-therm' consists of 2 virtual LED's – Abled1 and Abled2   whose physical characteristics- shape, size, and colour can be determined by software or by the Abled object inspector property panel.

Delphi Program LED1x

```
unit led1;
interface
uses
  Windows, Messages, SysUtils, Classes, Graphics, Controls, Forms, Dialogs,
  _Gclass, AbLED;
type
  Tform1 = class(Tform)
    AbLED1: TAbLED;
    AbLED2: TAbLED;
    procedure FormCreate(Sender: Tobject);
  private
    { Private declarations }
  public
    { Public declarations }
  end;
var
  Form1: Tform1;
implementation

{$R *.DFM}
procedure Tform1.FormCreate(Sender: Tobject);
begin
AbLED1.LED.Coloron:=clred;
AbLED1.LED.Coloroff:=clyellow;
AbLED2.LED.Coloroff:=clgreen;
AbLED2.LED.Coloron:=clblue;
AbLED1.LED.Height:=20;
AbLED2.LED.Height:=5;
end;
end.
```

Programming the virtual LED's.

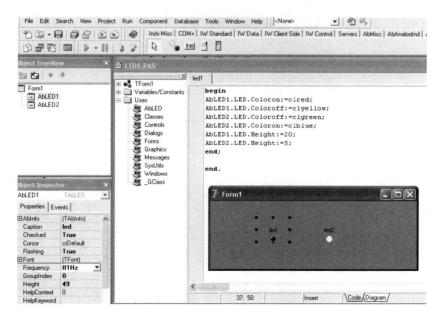

2.4 Delphi Tcolourbutton

The Tcolorbutton adds three new properties to the standard Delphi Tbutton and It can be downloaded from the http://delphi.about.com website and installed as a single new component.

The hovercolor specifies the colour used to paint the buttons background when the mouse pointer hovers over the button. The Forecolor specifies the colour of the button text and the Backcolor specifies the background colour of the button.

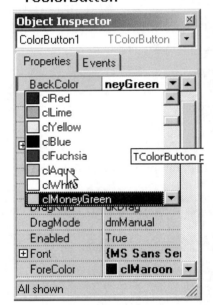

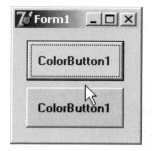
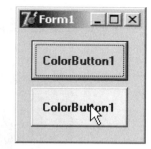

3 VCP USB communications

The next step to achieve virtual instrumentation is to provide a USB communications channel enabling the flow of digital data to and from the virtual components. FTDI's and Prolific RS232/USB bridges are ideally suited to provide a 'virtual communications port' or VCP . These devices are seen by the PC as a standard RS232 COM port although they connect to the USB port. Serial RS232 data can now be sent and received by an external device.

In keeping with all USB interfaces, these RS232/USB modules are no exception and require software drivers in order to function. This chapter shows how to install these drivers and detail the Case Study of a virtual thermometer. Scientific Instruments 'Portcontroller' Active X in conjunction with Delphi gives full control over the serial COM port with just a few lines of code.

3.1 Installing FTDI's VCP drivers

FTDI's EVAL232R shown below is a RS232/USB bridge with a standard USB and 9 pin D-type RS232 connector. The RS232 connector functions just like any RS232 port with TXD and RXD data lines.

EVAL232R FT232RL USB to RS232 Module

Connecting this module via a standard USB cable to the PC's port will invoke the following sequence of operations:

Click on 'No, not this time'.

Click on 'Install from a list or specific location (Advanced)'

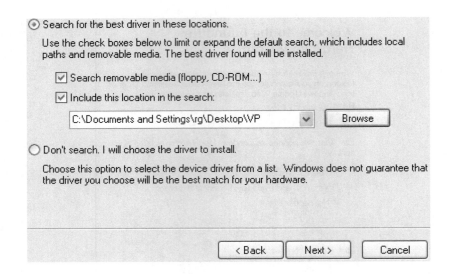

Click on 'Browse' and select the folder which contains the VCP driver.

Click on 'Finish'

The following message box will appear: Note that this time a USB Serial Port is found. The same menu options are presented as before.

Welcome to the Found New Hardware Wizard

Windows will search for current and updated software by looking on your computer, on the hardware installation CD, or on the Windows Update Web site (with your permission).
Read our privacy policy

Can Windows connect to Windows Update to search for software?

○ Yes, this time only
○ Yes, now and every time I connect a device
◉ No, not this time

Click Next to continue.

This wizard helps you install software for:

USB Serial Port

 If your hardware came with an installation CD or floppy disk, insert it now.

What do you want the wizard to do?

○ Install the software automatically (Recommended)
◉ Install from a list or specific location (Advanced)

Click Next to continue.

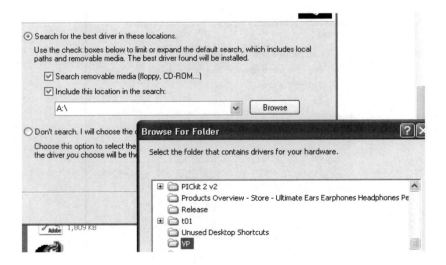

Completing the Found New Hardware Wizard

The wizard has finished installing the software for:

 USB Serial Port

Click Finish to close the wizard.

Now we have to locate which serial COM port has been assigned to the RS232/USB bridge. Select 'System' in Control Panel in Windows and click on 'Hardware', as below:

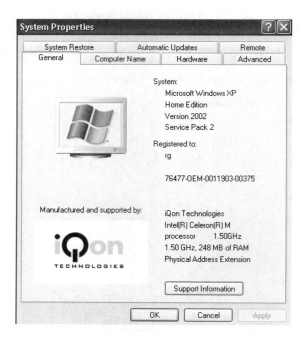

Click on 'Device Manager'

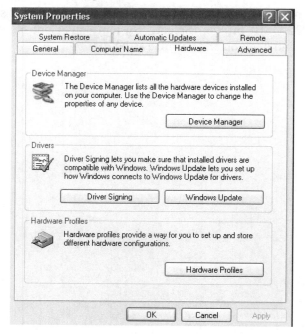

A list of devices is shown and it can be seen that serial port COM5 has been assigned to the USB bridge.

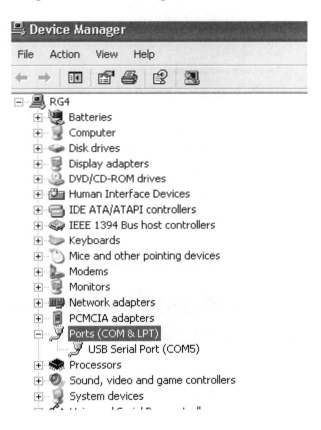

Clicking on 'USB Serial Port (COM5) ' and then on 'Port Settings' and 'Advanced' allows the user to change the port to for example COM port 1 (COM1) or COM2.

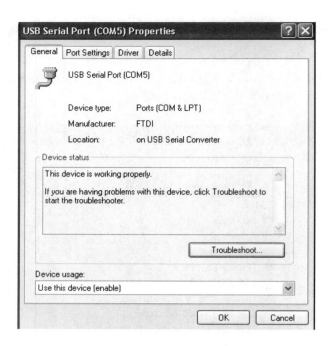

Select COM Port Number e.g. COM1 and click OK

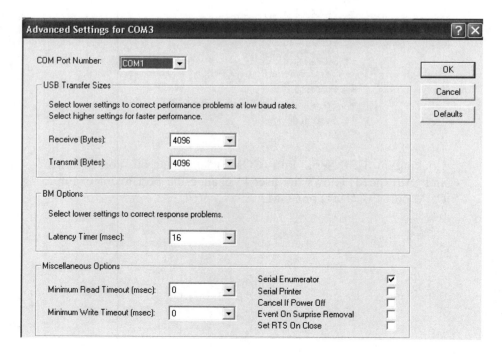

Now the USB bridge can be tested by a simple loopback connecting pins 2 and 3 on the 9 pin D-Type connector, connecting the USB cable from the bridge into the PC's USB port and running a terminal emulation program such as Windows 'Terminal' or 'Superterm'.

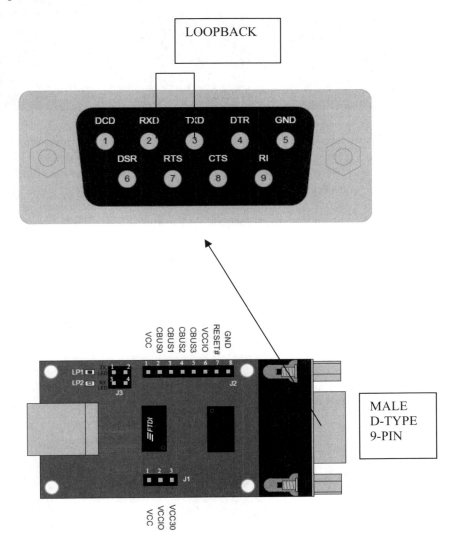

Simply pressing any key on the keyboard will cause that character to be echoed back via the loopback on the RS232 connector of the bridge.There is no need to set the baud rate, since the outgoing data and receiving data will always be at the same baud rate. The letters in red above are the characters pressed on the keyboard and the white letters, the character received from the USB/RS232 bridge.

3.2 RS232 connectors

D-Type 9 pin male and female RS232 connectors

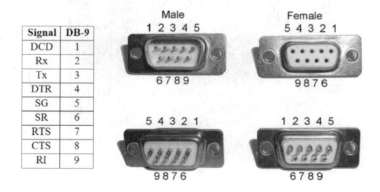

Signal	DB-9
DCD	1
Rx	2
Tx	3
DTR	4
SG	5
SR	6
RTS	7
CTS	8
RI	9

Led indicators can monitor the activity on the TXD and RXD lines of the RS232 port. For this, a converter and an RS232 tester is required as shown below:

9 pin to 25-pin converter   RS232 led Tester

Pin2 txd   Pin3 rxd   Pin7 gnd

A RS232 Tester connected to the EVAL232R module

4 pin and 25 pin gender changers

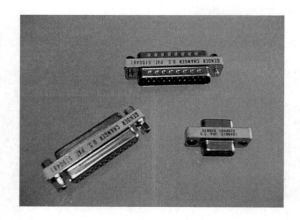

A selection of RS232 connectors and hoods

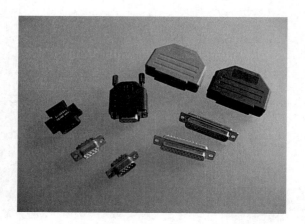

3.3 Portcontroller Active X

In order to receive data via a VCP USB port an Active X component is required. Scientific Instruments 'Portcontroller' simplifies RS232 serial communications and can be used in Delphi. Its library (see Appendix) contains all the instructions necessary to initialise the COM port and to import and export data via the USB port. Some of the Pascal functions and procedures are listed below:

```
function Get_BaudRate: Integer; safecall;
procedure Set_BaudRate(Value: Integer); safecall;
function Get_DataBits: Smallint; safecall;
procedure Set_DataBits(Value: Smallint); safecall;
function Get_Parity: Smallint; safecall;
procedure Set_Parity(Value: Smallint); safecall;
function Get_StopBits: Smallint; safecall;
procedure Set_StopBits(Value: Smallint); safecall;
function Get_PortName: WideString; safecall;
function Get_Cd: Integer; safecall;
function Get_Cts: Integer; safecall;
function Get_Dsr: Integer; safecall;
function Get_Dtr: Integer; safecall;
procedure Set_Dtr(Value: Integer); safecall;
function Get_DtrDsr: Integer; safecall;
procedure Set_DtrDsr(Value: Integer); safecall;
function Get_RtsCts: Integer; safecall;
procedure Set_RtsCts(Value: Integer); safecall;
function Get_Rts: Integer; safecall;
procedure Set_Rts(Value: Integer); safecall;
```

The trial software download can be found at:
http://www.scientificcomponent.com/
After installing Portcontroller ActiveX, start Delphi and choose Components / Import ActiveX Control from main menu.

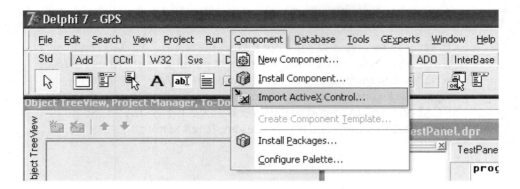

The following dialog appears. Scroll down to the PortController 2.0 entry and select it; Delphi will set the other parameters automatically. Click the Install button.

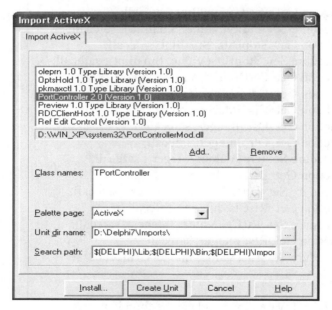

Select Into new package tab page and enter the name of the package file into which you want to install Portcontroller. You may also type description for this package. Click the OK button.

After confirming the message dialogs the Portcontroller package will appear.

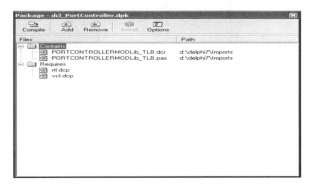

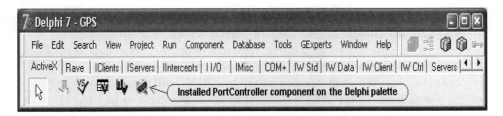

Scroll the Delphi palette to the ActiveX page and confirm that Portcontroller exists on the page.

You may close the Package window. You can create a Portcontroller object by dragging the Portcontroller icon onto designed form. You may now begin coding with Portcontroller. See the Portcontroller Reference at the end of the book for details on how to use Portcontroller properties, methods, and events.

Delphi Program with Active X component installed.

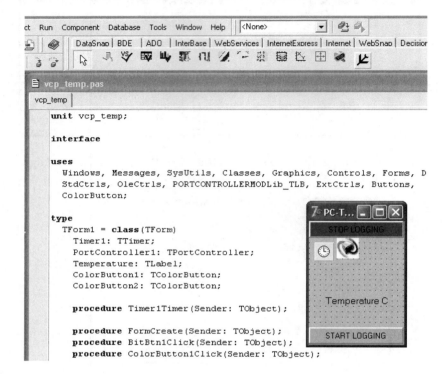

Delphi program VIX uses Portcontroller to import a Byte via a RS232/USB convertor and display on the screen. The Byte, in this case 4EH, can represent an ASCII text character (N) or data. For virtual instrumentation the byte represents a real value –for example, for a 5 volt voltage range, an ADC will produce a byte between 0 and FFh (255).We can then scale this by dividing by 51 so that the displayed result will show a value between 0 and 5 volts i.e. 78 decimal /51 = 1.53.

DATA and TEXT representation of a Byte

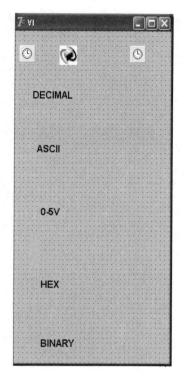 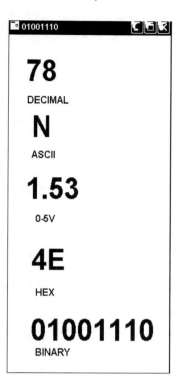

Delphi Project VIX
unit vi;

interface

uses
  Windows, Messages, SysUtils, Classes, Graphics, Controls, Forms, Dialogs,
  StdCtrls, OleCtrls, PORTCONTROLLERMODLib_TLB, ExtCtrls, Buttons,
  ColorButton, ActnList;

type
  Tform1 = class(Tform)
    Timer1: Ttimer;
    PortController1: TportController;
    Label1: Tlabel;
    Label2: Tlabel;
    Label3: Tlabel;
    Label4: Tlabel;
    Label5: Tlabel;
    Timer2: Ttimer;
    Label6: Tlabel;

```
procedure Timer1Timer(Sender: Tobject);

procedure FormCreate(Sender: Tobject);

procedure Timer2Timer(Sender: Tobject);

private
  { Private declarations }
public
  { Public declarations }
end;


var
  Form1: Tform1;

    s:string;

numberbytesread:Integer;
data:string[1];
x:byte;
a:real;

   Buf: array[0..SizeOf(Extended) * 2] of Char;

implementation

{$R *.DFM}


procedure Tform1.Timer1Timer(Sender: Tobject);

begin
  Timer1.Enabled:=False;
PortController1.ClearRQ;
data:=PortController1.Read(1, 0,numberbytesread);
asm
mov ax,data[1];
mov x,al
end;


a:=x;

Canvas.TextOut(30,30,'              ');
Canvas.TextOut(30,120,'             ');
Canvas.TextOut(30,220,'             ');

Canvas.TextOut(40,120,data);   {ASCII}
Str(a:1:0,s);
Canvas.TextOut(30,30,s);       {DECIMAL}
a:=a/51;
Str(a:2:2,s);                  {SCALED}
Canvas.TextOut(30,220,s);
```

```
    BinToHex(@data, Buf, 2);      {HEX}

   Label3.Caption := Format(' %s', [Buf+2]);
   Timer1.Enabled:=True;
   end;

procedure Tform1.FormCreate(Sender: Tobject);
begin
 Timer1.Enabled:=True;
  Timer1.Interval:=1;
   PortController1.Open('COM3','9600,8,N,1');

PortController1.RtsCts:=Ord(true);
  PortController1.DtrDsr :=Ord(true);
end;


Function HexToBin(Hexadecimal: string): string;
const
  BCD: array [0..15] of string =
    ('0000', '0001', '0010', '0011', '0100', '0101', '0110', '0111',
    '1000', '1001', '1010', '1011', '1100', '1101', '1110', '1111');
var
 i: integer;

begin
 for i := Length(Hexadecimal) downto 1 do

    Result := BCD[StrToInt('$' + Hexadecimal[i])] + Result;
end;

procedure Tform1.Timer2Timer(Sender: Tobject);
begin
Caption:=(HexToBin(Buf+2));     {BINARY}
Canvas.TextOut(40,450,Caption);
end;

end.
```

3.4 Case study VCP DS75 virtual digital and analogue thermometer

Digital Thermometer readout

Our first virtual instrument requires three elements – a USB/RS232 bridge, a PIC controlled temperature sensor and a Delphi virtual PC digital thermometer. The sensor used is a DS75 digital thermometer and thermostat which provides 9,10,11 or 12-bit digital temperature readings over a –55 degree C to +125 degree C. Communications between it and a PIC12F629 is achieved via a simple 2-wire serial interface. Three address pins allow up to eight DS75 devices to operate on the same 2-wire bus, which greatly simplifies distributed temperature sensing applications.

The DS75

PIN ASSIGNMENT

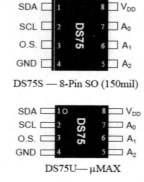

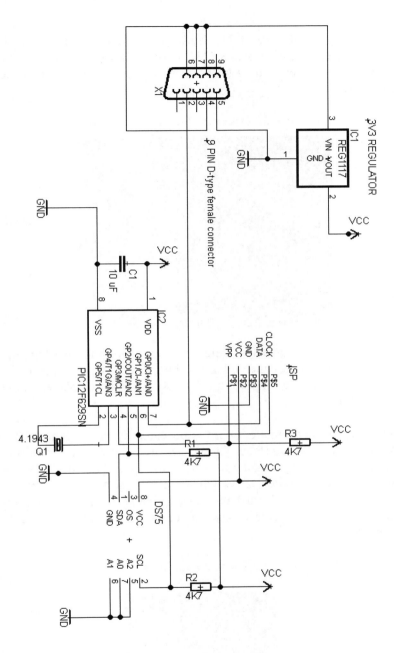

Schematic diagram Digital Thermometer

In this cicuit, FTDI's EVAL232 bridge is used configured as a VCP device and connected via a RS232 D-type connector. Indeed any RS232/USB convertor may be used including the Prolific and the Belkin bridge.

Prolific and Belkin RS232/USB converters

The Thermometer displays a digital temperature and also logs the temperature at 1 second intervals into file 'Temperature.TXT' which then can be imported into EXCEL and displayed as a graphical trend.

Power to the Thermometer board is provided by the RTS and DTR lines of the RS232 interface. These lines are connected together and fed to a 3V3 voltage regulator. This helps sync with the PC since the board only becomes active when the PC's software enables the RTS/DTR hardware handshake lines. Finally an ISP (in circuit programming) connector is used to connect to a PICKIT 2 PIC programmer.

The Thermometer board

The DS75 device stores the temperature in a 2 byte internal register. The high byte holds the temperature value of + 0 to 125 degrees, the low byte holds fractional temperature values. Below zero degrees C the value is stored as a 2's compliment number. The table below shows the register values for several different temperatures.

Figure 3. TEMPERATURE, T_H, and T_L REGISTER FORMAT

	bit 15	bit 14	bit 13	bit 12	bit 11	bit 10	bit 9	bit 8
MS Byte	S	2^6	2^5	2^4	2^3	2^2	2^1	2^0
	bit 7	bit 6	bit 5	bit 4	bit 3	bit 2	bit 1	bit 0
LS Byte	2^{-1}	2^{-2}	2^{-3}	2^{-4}	0	0	0	0

Table 3. 12-BIT RESOLUTION TEMPERATURE/DATA RELATIONSHIP

TEMPERATURE (°C)	DIGITAL OUTPUT (BINARY)	DIGITAL OUTPUT (HEX)
+125	0111 1101 0000 0000	7D00h
+25.0625	0001 1001 0001 0000	1910h
+10.125	0000 1010 0010 0000	0A20h
+0.5	0000 0000 1000 0000	0080h
0	0000 0000 0000 0000	0000h
-0.5	1111 1111 1000 0000	FF80h
-10.125	1111 0101 1110 0000	F5E0h
-25.0625	1110 0110 1111 0000	E6F0h
-55	1100 1001 0000 0000	C900h

For example a +ve temperature of 21.75 degrees would be stored as:

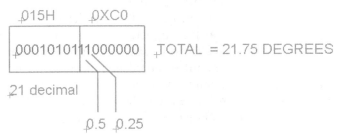

For a temperature below zero, (bit 15 of the data will be a logic 1) the registers for a temperature of – 10.125 degrees shown below, a 2's complement operation has to be performed on the data to extract the temperature value. It was decided to do this not in the PIC software but in the Delphi program.

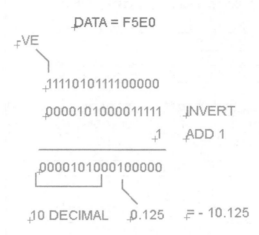

The PIC software- sensor33-2400.ASM and the HEX code for the digital thermometer can be downloaded from the ELEKTOR.COM Website and can be found in Folder vcp_prolific_ftdi_temp.

The Delphi Project (vcp_ftdi_temp) with file vcp_temp.pas runs on a PC and is a software program that replicates the image and functionality of a digital thermometer. It does not use any visual instrument components and only uses the colorbutton component as buttons. Once the thermometer board is connected to the PC via a standard USB lead, running the program and clicking on the start button starts operation. To test the thermometer, freezer spray was applied to the sensor and then a soldering iron tip was applied for a few seconds. The results are shown below: The X axis calibrated in degrees C, The Y axis number of samples at 1 sample/sec.

Freezer and heat test for the thermometer.

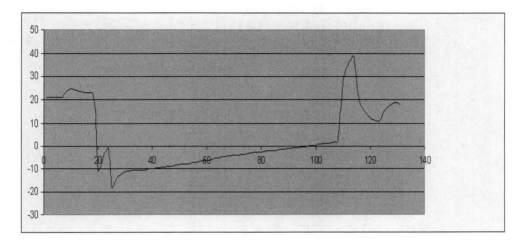

To create the graph, open file temperature .TXT in EXCEL, click on the A column which will automatically highlight all the data values in the column and then, click on the chart symbol and select the type of graph. The X axis is calibrated in degrees C and the Y axis in number of samples.

Importing logged temperature data into EXCEL

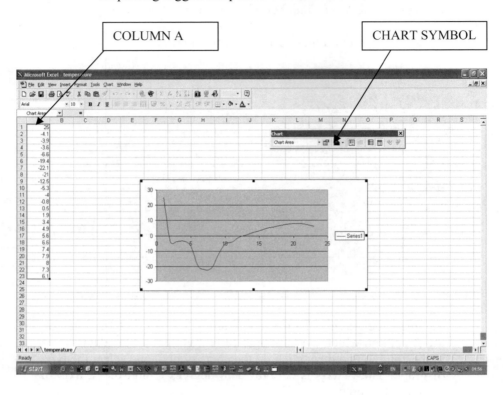

The Delphi Project 'vcp_ftdi_temp' for the virtual digital thermometer in Folder vcp_prolific_ftdi_temp.:

(a)

```
unit vcp_temp;

interface

uses
  Windows, Messages, SysUtils,
Classes, Graphics, Controls, Forms,
Dialogs,
  StdCtrls, OleCtrls,
PORTCONTROLLERMODLib_TLB,
ExtCtrls, Buttons,
  ColorButton;

type
  TForm1 = class(TForm)
    Timer1: TTimer;
    PortController1: TPortController;
    Temperature: TLabel;
    ColorButton1: TColorButton;
    ColorButton2: TColorButton;

    procedure Timer1Timer(Sender:
TObject);

    procedure FormCreate(Sender:
TObject);
    procedure BitBtn1Click(Sender:
TObject);
    procedure
ColorButton1Click(Sender: TObject);
    procedure
ColorButton2Click(Sender: TObject);
  private
    { Private declarations }
  public
    { Public declarations }
  end;



var
  Form1: TForm1;

    s:string;
```

(b)

```
numberbytesread:Integer;
data:string[1];
x:word;
a:real;
msb:byte;
lsb:byte;
n:byte;
samples:integer;
txt:text;
flag:boolean;
timestr:string;
implementation

{$R *.DFM}


procedure
TForm1.Timer1Timer(Sender:
TObject);
LABEL OUT;
begin
Timer1.Enabled:=False;
flag:=true;
n:=0;
data:=PortController1.Read(1,
0,numberbytesread);
{data:='!'; }
asm
push ax
mov ax,data[1];
mov msb,al;
pop ax
end;
data:=PortController1.Read(1,
0,numberbytesread);
{data:='!';
```

91

(c)
asm

push ax
push bx

mov ax,data[1];
mov lsb,al;
mov ah,msb;
CMP AH,0;
JNs @OUT
XOR AX,0ffffh
ADD AX,1
mov bl,1
mov n,bl
@OUT:
shr ax,1;
shr ax,1;
shr ax,1;
shr ax,1;
shr ax,1;
mov x,ax;

pop bx
pop ax
end;

a:=x;
a:=a/8;
timestr:=TimeToStr(Time);
{write(txt,' ',timestr,' '); }
if n=1 then
write(txt,'-');
writeln(txt,a:2:1);
Str(a:2:1,s);
Caption:=TimeToStr(Time);
Canvas.TextOut(10,30,'   ');
if n=1 then
Canvas.TextOut(10,29,'-');
Canvas.TextOut(30,30,s);
flag:=false;
 Timer1.Enabled:=True;
end;

(d)

procedure TForm1.FormCreate(Sender: TObject);
begin
Timer1.Enabled:=False;
 Timer1.Interval:=10;
PortController1.RtsCts:=Ord(true);
PortController1.DtrDsr:=Ord(true);
AssignFile(txt,'C:\remote.txt');
Rewrite(txt);
flag:=false;
end;

procedure TForm1.BitBtn1Click(Sender: TObject);
begin
Timer1.Enabled:=False;
end;

procedure TForm1.ColorButton1Click(Sender: TObject);
begin
{while flag= false do }
ColorButton1.Enabled := False;
if flag=false then
Timer1.Enabled:=False;
PortController1.Close;
CloseFile(txt);
close;
end;

procedure TForm1.ColorButton2Click(Sender: TObject);
begin

PortController1.Open('COM2','2400, 8,N,1');
Timer1.Enabled:=tRUE;
ColorButton2.Enabled := False;
end;
end.

We can now introduce the Abacus Abthermometer virtual component, shown below, into the program to create an analogue thermometer which uses the identical thermometer PIC-based hardware board as for the digital thermometer.

The analogue Thermometer

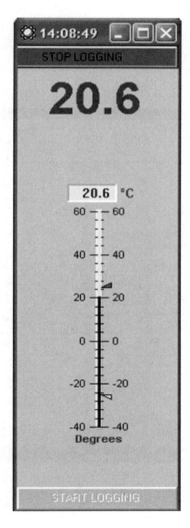 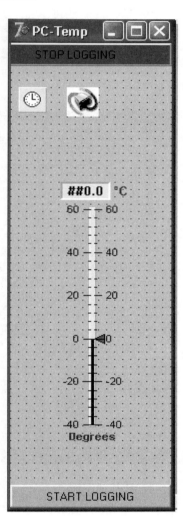

Executable   Delphi Form

The Delphi project 'vcp_ftdi_temp_analogue' for the analogue thermometer can be found in the vcp_prolific_ftdi_temp folder.

A single Byte variable (a) in statement 'Abthermometer1.Value:=a;' is used to control the thermometer movement and is updated every second. COM port 2 is used with a baud rate of 2400 to match the output from the PIC temperature probe.

Room Temperature graph

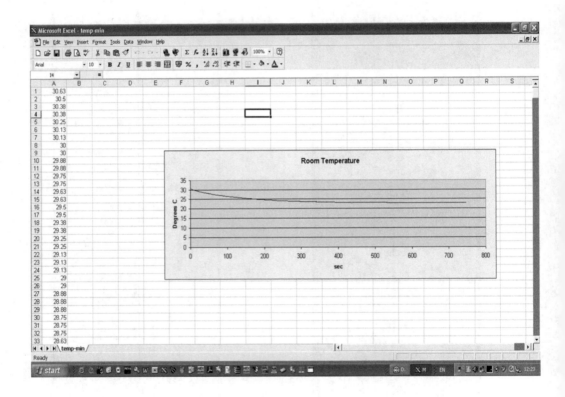

4 DLL USB communications

FTDI's royalty-free DLL driver enables FTDI's products to function as a DLL device and to dispense for the need for virtual RS232 COM ports. A DLL library – D2XXUnit (see Appendix) is provided by FTDI to be used in the DELPHI environment to enable USB communications. The USBOUT LED test board case study shows how data can be sent from the PC's USB port using DLL serial communications. Three different virtual control panels are used to output data to this test board. Also in this chapter is a case study demonstrating Microchip's 44 pin demo board with an on-board potentiometer which is used to control a virtual meter.

4.1 Installing FTDI's DLL driver

Before a DLL driver can be installed on the PC, the VCP driver has to be removed with the aid of FTDI's FTClean program which can be downloaded from the FTDI.COM website. Disconnect the USB lead, uncompress the file, run Ftclean.exe, click 'clean system' and then click YES to confirm.

The FTClean program

DLL driver installation is identical to that of installing a VCP driver except that there is only one iteration. The USB lead can now be reconnected to the FTDI device. The installation wizard will run and the DLL driver will be installed.

This wizard helps you install software for:

USB <-> Serial Cable

 If your hardware came with an installation CD or floppy disk, insert it now.

What do you want the wizard to do?

○ Install the software automatically (Recommended)
◉ Install from a list or specific location (Advanced)

Click Next to continue.

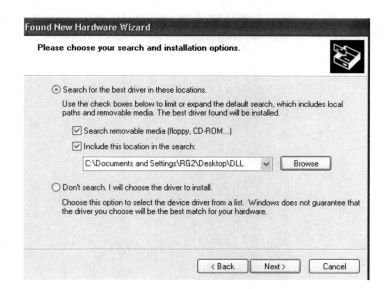

97

Completing the Found New Hardware Wizard

The wizard has finished installing the software for:

 FTDI FT8U2XX Device

Click Finish to close the wizard.

To see if the DLL driver has been successfully installed, 'USBView' is used which is a program that displays the topography of devices that are plugged into the USB bus. It can be seen that the FTDIFT8U2XX device is installed:

The USBView program

If we now go into the Device manager the Universal Serial Bus controller list now contains the FTDI FT8U2XX Device:

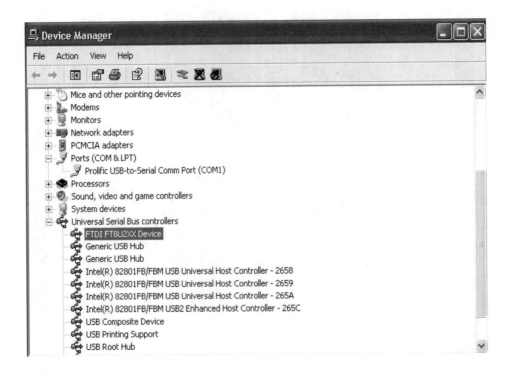

USB connectors

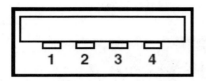 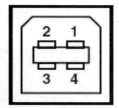

1 +5 V
2 Data –
3 Data +
4 Ground

4.2 Case Study DLL LED test board

This LED output board is based on FTDI's parallel DLP-USB245M USB/RS232 convertor. Data is still passed from the USB port as 8-bit data but presented on an 8 bit wide parallel bus. A PIC12F629 controls the latching of the data from D0-D7 lines of the convertor onto a 74HC573 octal D type transparent latch which in turn switches 8 LED's. For example, a byte 081H (10000001) binary, received by the convertor will switch on LED1 and LED8. An LED indicator LED0 flashes at 1 second intervals conforming that the board is connected to the USB port and is operational.

DLL USBOUT LED test board

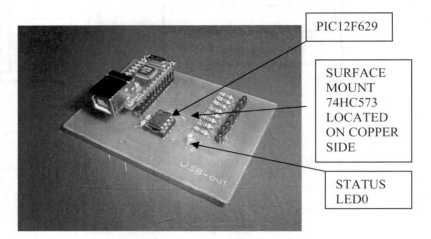

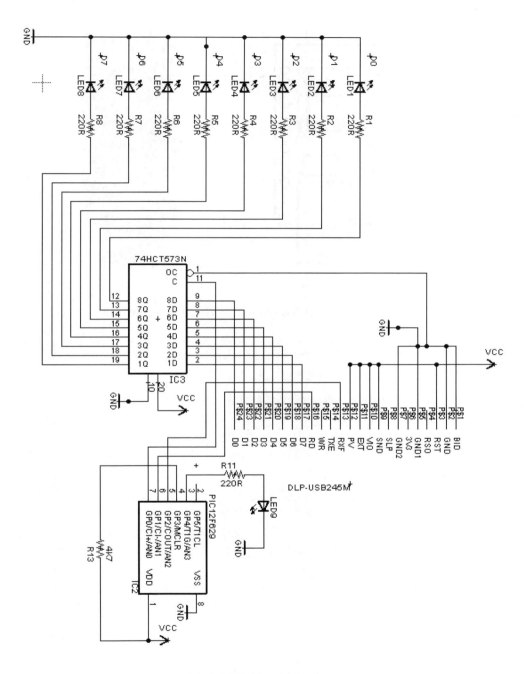

DLL LED Test Board

Table 1 - DLP-USB245M PINOUT DESCRIPTION

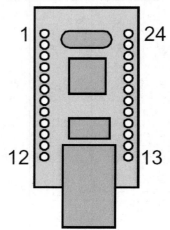 

The DLP-USB245M parallel RS232/USB convertor

Pin#	Description
1	BOARD ID (Out) Identifies the board as either a DLP-USB245M or DLP-USB232M. High for DLP-USB232M and low for DLP-USB245M.
2	Ground
3	RESET# (In) Can be used by an external device to reset the FT245BM. If not required this pin must be tied to VCC.
4	RESETO# (Out) Output of the internal Reset Generator. Stays high impedance for ~ 2ms after VCC > 3.5v and the internal clock starts up, then clamps it's output to the 3.3v output of the internal regulator. Taking RESET# low will also force RSTOUT# to go high impedance. RSTOUT# is NOT affected by a USB Bus Reset.
5	Ground
6	3V3OUT (Out) Output from the integrated L.D.O. regulator. It's primary purpose is to provide the internal 3.3v supply to the USB transceiver cell and the RSTOUT# pin. A small amount of current (<= 5mA) can be drawn from this pin to power external 3.3v logic if required.
7	Ground
8	SLEEP (Out) Goes Low after the device is configured via USB, then high during USB suspend. Can be used to control power to external logic using a P-Channel Logic Level MOSFET switch.

9	SND/WUP (In) If the DLP-USB245M is in USB suspend, a positive edge on this pin (WAKEUP) initiates a remote wakeup sequence. If the device is active (not in suspend) a positive edge on this pin (SEND) causes the data in the write buffer to be sent to the PC on the next USB Data-In request regardless of how many bytes are in the buffer.
10	VCC-IO (In) 3.0 volt to +5.25 volt VCC to the UART interface pins 10..12, 14..16 and 18..25. When interfacing with 3.3v external logic connect VCC-IO to the 3.3v supply of the external logic, otherwise connect to VCC to drive out at 5v CMOS level. This pin must be connected to VCC from the target electronics or EXTVCC.
11	EXTVCC – (In) Use for applying main power (4.4 to 5.25 Volts) to the module. Connect to PORTVCC if module is to be powered by the USB port (typical configuration)
12	PORTVCC - (Out) Power from USB port. Connect to EXTVCC if module is to be powered by the USB port (typical configuration). 500mA maximum current available to USB adapter and target electronics if USB device is configured for high power.
13	RXF# - (Out) When low, at least 1 byte is present in the FIFO's 128-byte receive buffer and is ready to be read with RD#. RXF# goes high when the receive buffer is empty.
14	TXE# - When high, the FIFO's 385 byte transmit buffer is full, or busy storing the last byte written. Do not attempt to write data to the transmit buffer when TXE# is high.
15	WR (In) When taken from a high to a low state, WR reads the 8 data lines and writes the byte into the FIFO's transmit buffer. Data written to the transmit buffer is sent to the host PC within the TX buffer timeout value (default 16mS) and placed in the RS-232 buffer opened by the application program. Note : The FT245BM allows the TX buffer timeout value to be reprogrammed to a value between 1 and 255mS depending on the applicaton requirement, also the SND pin can be used to send any remaining data in the TX buffer regardless of the timeout value.
16	RD# (In) When pulled low, RD# takes the 8 data lines from a high impedance state to the current byte in the FIFO's receive buffer. Taking RD# high returns the data pins to a high impedance state and prepares the next byte (if available) in the FIFO to be read.
17	D7 I/O Bi-directional Data Bus Bit # 7
18	D6 I/O Bi-directional Data Bus Bit # 6
19	D5 I/O Bi-directional Data Bus Bit # 5
20	D4 I/O Bi-directional Data Bus Bit # 4
21	D3 I/O Bi-directional Data Bus Bit # 3
22	D2 I/O Bi-directional Data Bus Bit # 2
23	D1 I/O Bi-directional Data Bus Bit # 1
24	D0 I/O Bi-directional Data Bus Bit # 0

The source code for the PIC12F629 can be found in Folder USB-OUT and is listed below:

Program USBOUT.ASM

```
list    p=12F629    ; list directive to define processor
        #include <p12f629.inc>   ; processor specific variable definitions

        __CONFIG _CP_OFF & _WDT_OFF & _BODEN_ON & _PWRTE_ON &
_INTRC_OSC_NOCLKOUT & _MCLRE_ON & _CPD_OFF

;AUTHOR: Richard Grodzik July 15th 2005
;*************************************************************************
;Defines
;*************************************************************************

#define Bank0           0x00
#define Bank1           0x80
#define SWITCH          GPIO,3
#define D0_1Tris    B'00000010'
#define D0On            B'00010000'
#define D0Off           B'00000000'
#define LEDOn           Flags,0


RTCC    EQU 1       ;RTCC
PC      EQU 2
STATUS  EQU 3       ;STATUS REGISTER
BUFFER EQU 020H
COUNTR EQU 021H
COUNTER EQU 022H
DLYCNT EQU 023H
THIRTYTWO EQU 024H
SIGNALS EQU 025H
GPIO   EQU 5
CMCON EQU 019H
ANSEL EQU 09FH

        ORG    0x000
    goto begin


        ORG 0x004 ;INTERRUPT VECTOR FOR GPIO,2
        bcf INTCON,1
        bcf INTCON,7 ;DISABLE GLOBAL INTERRUPT

; INCLUDED THESE LINES TO POLL INTERRUPT LINE FOR LOW
;....................................
;WAIT   BTFSC GPIO,2 ;POLL INTERRUPT LINE
;           GOTO WAIT
;....................................
  BSF GPIO,4
```

```
            BSF GPIO,0   ;ALE LINE for 74HC573 LATCH TRANSPARENT
            BCF GPIO,1   ;ISSUE READ LINE ACTIVE LOW

    BCF GPIO,0   ;ALE LATCHED DATA D0-D7
    BSF GPIO,1   ;BRING READ LINE BACK HIGH


     BSF GPIO,4  ;LED INDICATOR THAT DATA HAS RECEIVED
    ; NOP
    ; BCF GPIO,4

            bsf INTCON,7 ;ENABLE GLOBAL INTERRUPT
            bcf INTCON,1 ;CLEAR INTERRUPT FLAG
            retfie       ;RETURN FROM INTERRUPT

begin

            nop

            ; required by in circuit debugger
            BCF STATUS,RP0   ;BANK 0
                ;INIT GPIO

            MOVLW 7
            MOVWF CMCON       ;DISABLE COMPARATOR
                CLRF GPIO

    MOVLW B'11010000'         ; B'11010000'
            MOVWF INTCON

            BSF STATUS,RP0              ; BANK1
            ;movwf   OSCCAL

    MOVLW B'00100'
            MOVWF TRISIO      ; update register with factory cal value
            CLRF ANSEL        ;DIGITAL IO

            ;BSF IOC,2             ;

    MOVLW B'00000111';  PRESCALER 00000111 (/256)
    MOVWF  OPTION_REG
            ;call   0x3FF     ; retrieve factory calibration value
                                                       ; comment instruction if
using simulator, ICD2, or ICE2000

            BCF STATUS,RP0            ; change back to PORT memory bank



;***********************************************************************
***
;Main

;***********************************************************************
***
```

```
            BSF GPIO,0   ;ALE  HIGH
            BSF GPIO,1   ;READ HIGH
            BSF GPIO,4   ;LED
            CALL SECOND
            BCF GPIO,4

BUG NOP
                     BSF GPIO,4   ;LED
            CALL SECOND
            BCF GPIO,4
  CALL SECOND
            GOTO BUG


SECOND

   CLRF RTCC
   CLRWDT


   MOVLW 8
   MOVWF THIRTYTWO  ;COUNTER
   CLRF RTCC       ;LOAD RTCC

SHORT BTFSC RTCC,7   ;TEST FOR 128 DECIMAL
            ;I.E. BIT 7 = HIGH
   GOTO JMP2
   CLRWDT
   GOTO SHORT
JMP2 CLRF RTCC
            ;32 LOOP COUNTER
  DECFSZ THIRTYTWO,1
  GOTO SHORT

  RETURN
DELAY
            MOVLW .10;255
            MOVWF THIRTYTWO
HERE NOP
            NOP
            NOP
            NOP
            NOP
            NOP
            NOP
            NOP
            NOP
            NOP
            DECFSZ THIRTYTWO,1
            GOTO HERE
            RETURN

     END
```

4.3 Simple DAC for 0-5 volt output

A simple DAC board for the LED test board utilising a DAC0800N is shown below. The 8 data lines of the DAC0800N (D0-D7) are connected to the 1Q-8Q outputs of the 74HC573 latch. The incoming byte to the LED board will now produce a D.C. output voltage on JP2 varying from 0 to 5 volts dependant on the value of the data. A 7660 produces a balanced supply of + and – 5 volts for the DAC.

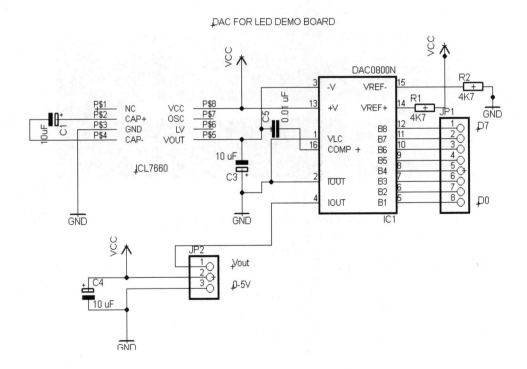

4.4 Delphi VI slider control panel for test board

A simple scroll bar can now be used to test the LED test board.
Delphi project 'NumberxDLL' in folder NUMBERDLL uses FTDI's D2XX unit's procedures and functions (found in the appendix at the end of the book) to initialise the serial RS232 communications and to write data to the LED test board. Moving the scroll bar from left to right outputs data from 0 to 255 binary. i.e. with the scroll bar at the limit of it's travel, all 8 LED's will be lit. A binary representation of the data will be shown on the LED's for any position of the scroll bar, together with a d.c. voltage on the DAC output proportional to the scroll bar position. In the example shown below, 24H (36 decimal) will be sent to the test board causing LED's D2 and D6 to be lit.

The delphi program for the slider control is listed below:
Program Numberxxdll.pas

```
unit numberxxDLL;

interface

uses
  Windows, Messages, SysUtils, Classes, Graphics, Controls, Forms, Dialogs,
  StdCtrls, ColorButton, ExtCtrls;

type
  TForm1 = class(TForm)
    Timer1: TTimer;
    ScrollBar1: TScrollBar;
    procedure FormCreate(Sender: TObject);
    procedure ScrollBar1Change(Sender: TObject);
  private
    { Private declarations }
```

```
  public
    { Public declarations }
  end;

var
  Form1: TForm1;
  FT_Current_Baud:Dword;

  I:Integer;
  txt:text;
   n:integer;
    v:Byte;
    DevicePresent : Boolean;
 Selected_Device_Serial_Number : String;
 Selected_Device_Description : String;



const

    FT_DLL_NAME='ftd2xx.dll';
    FT_Out_Buffer_Size=1;
Implementation'[\;
uses D2XXUnit;
{$R *.DFM}

procedure TForm1.FormCreate(Sender: TObject);
begin
 Open_USB_Device;

  FT_Enable_Error_Report := false; // Disable Error Reporting
  Timer1.Enabled := False;  // Stop Polling for Device Present

n:=0;
Canvas.Pen.Color:=ClLime;


FT_Current_Baud:=FT_BAUD_9600;
FT_Current_DataBits:=FT_DATA_BITS_8;
FT_Current_StopBits:=FT_STOP_BITS_1;
FT_Current_Parity:=FT_PARITY_NONE;
 Set_USB_Device_BaudRate;

 FT_Out_Buffer[0]:=0;
I := Write_USB_Device_Buffer(FT_Out_Buffer_Size);


end;

procedure TForm1.ScrollBar1Change(Sender: TObject);

var
   Optval:Byte;
 SaveFile : File;
   BytesWrote,Total,I : Integer;
   SaveFileName : String;
```

```
    PortStatus : FT_Result;
    s:string;
    B:Byte;
     x:integer;
     n:integer;


begin

Optval:=Scrollbar1.Position;

Str (optval,s);
Caption:=s;
 I:=1;
  Total:=1;
b:= StrToInt(s);


 FT_Out_Buffer[0]:=b;
I := Write_USB_Device_Buffer(FT_Out_Buffer_Size);


end;
end.
```

4.5 Delphi VI button control panel for test board

Delphi project 'Paneldllx2 'shown below in folder paneldllx shows how individual LED's can be controlled on the LED test board. In addition 2 buttons – OFF and ON will turn off/on all the LED's. The colorbutton component is used for the buttons which change colour when activated.

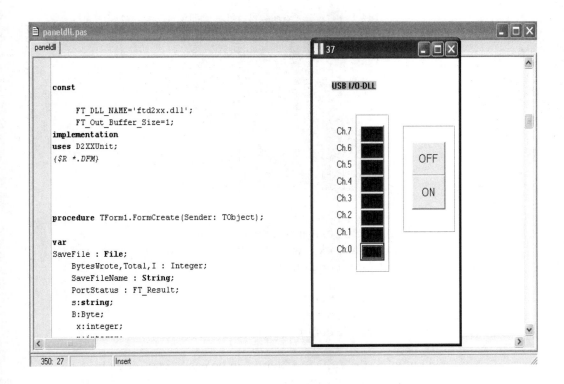

4.6 Delphi VI control panel for LED test board

A more sophisticated virtual instrument control panel may now be constructed using 'Uniworks' virtual components. Shown above the control panel uses three components:

( )UniguageX1 – a moving coil meter
( )UniLCDX1 – a digital meter
( )UniknobX1 – a simple control knob.

Now, the output to the LED test board can be easily controlled by a simple control knob, with the meter moving in sympathy with the knob and after scaling, the equivalent voltage produced on the DAC is displayed on the digital LCD meter.

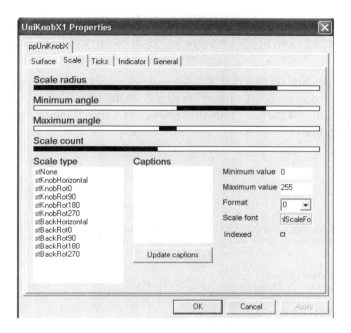

By right clicking on the component and selecting 'properties', the appearance of each component can be infinately modified, together with the scale, ticks and minimum and maximum values.

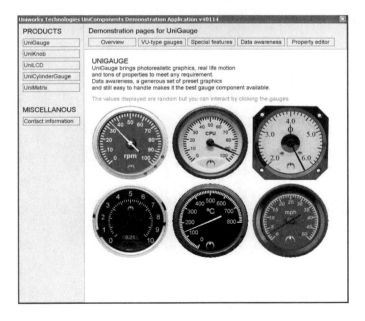

A trial software virtual instrumentation component library can be downloaded from http://www.uniworkstech.com/download.htm

Delphi Project 'Project11' for the control panel for LED testboard can be downloaded from Folder 'Paneluniworks' from the ELEKTOR website. Shown below is the Delphi program UNIT11.PAS

```
unit Unit11;

interface

uses
  Windows, Messages, SysUtils, Classes, Graphics, Controls, Forms, Dialogs,
  StdCtrls, OleCtrls, UniComponentsCOM_TLB, DBOleCtl, ExtCtrls;

type
  TForm1 = class(TForm)
    UniGaugeX1: TUniGaugeX;
    UniLCDX1: TUniLCDX;
    UniKnobX1: TUniKnobX;
    Timer1: TTimer;
    procedure Timer1Timer(Sender: TObject);
    procedure FormCreate(Sender: TObject);
    procedure ScrollBar1Change(Sender: TObject);
  private
    { Private declarations }
  public
    { Public declarations }
  end;

var
  Form1: TForm1;
    FT_Current_Baud:Dword;



    Optval:byte;
SaveFile : File;
    BytesWrote,Total,I : Integer;
    SaveFileName : String;

  s:string;
  B:Byte;
   x:integer;
   n:integer;



    DevicePresent : Boolean;
  Selected_Device_Serial_Number : String;
  Selected_Device_Description : String;
```

```
const

   FT_DLL_NAME='ftd2xx.dll';
   FT_Out_Buffer_Size=1;
implementation
uses D2XXUnit;
{$R *.DFM}


procedure TForm1.Timer1Timer(Sender: TObject);
var ii:real;
begin
ii:= UniKnobX1.Float;
UniGaugeX1.Float:=ii;
Optval:=round(ii);
{Optval:=Scrollbar1.Position; }
  UniLCDX1.Value:=Optval/51;
Str (optval,s);
Caption:=s;
 I:=1;
 Total:=1;
b:= StrToInt(s);

 FT_Out_Buffer[0]:=b;
I := Write_USB_Device_Buffer(FT_Out_Buffer_Size);

end;

 procedure TForm1.FormCreate(Sender: TObject);
 begin
    Open_USB_Device;
  FT_Enable_Error_Report := false; // Disable Error Reporting

n:=0;
Canvas.Pen.Color:=ClLime;

FT_Current_Baud:=FT_BAUD_9600;
FT_Current_DataBits:=FT_DATA_BITS_8;
FT_Current_StopBits:=FT_STOP_BITS_1;
FT_Current_Parity:=FT_PARITY_NONE;
 Set_USB_Device_BaudRate;

 FT_Out_Buffer[0]:=0;
I := Write_USB_Device_Buffer(FT_Out_Buffer_Size);
end;

procedure TForm1.ScrollBar1Change(Sender: TObject);
  var
  PortStatus : FT_Result;
begin
end;

end.
```

4.7 Delphi VI gauge for PICKIT 2 44 pin microchip demo board

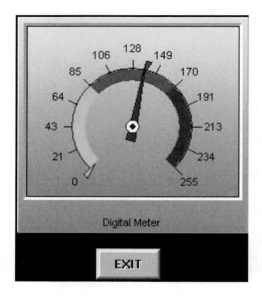

This time we shall use a virtual instrument as an input device displaying data that comes into the USB port. The EVAL232R FTDI RS232/USB bridge is used with a DLL driver. The source of the data is Microchip's 44 pin demo board which is populated with a PIC16F877 i/pt microcontroller and a potentiometer connected to an ADC port line. Program A2D for the PIC16F877 uses RA0 as an ADC input for the potentiometer and converts the result to an RS232 Byte using the 'bit-bang' method. The PIC's internal USART is not used since it generates NRZI format. The converted data is fed to the RXD pin on the EVAL232R bridge from port line RC6. Power is provided by connecting in a PICkit programmer to the ISP port of the demo board. As the potentiometer is turned the virtual digital meter displays the inbound data. The ticks on the display can be easily changed in the Delphi program to show a full scale voltage of 5 volts corresponding to the potentiometers maximum position. Note that a 20 Mhz crystal is fitted to the demo board to provide a baud rate of 9600.

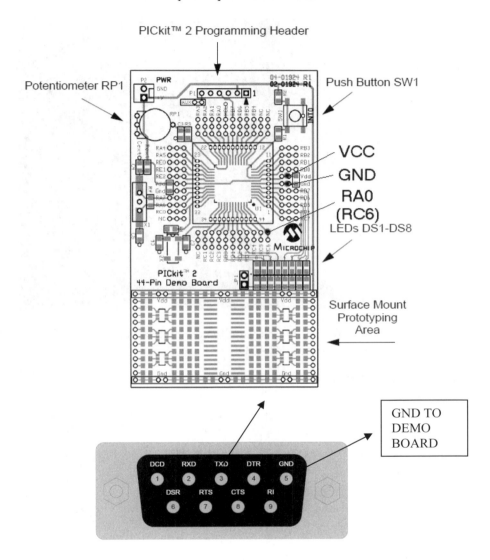

FTDI EVAL232R USB Bridge connections to Demo board

Microchip 44 pin Demo Board with PICkit2 connected

The program for the PIC is listed below and can be found in Folder 04 A2D+potdll, subfolder A2D.

Program A2D.ASM

```
#include <p16F887.inc>
        __CONFIG   _CONFIG1, _LVP_OFF & _FCMEN_OFF & _IESO_OFF &
_BOR_OFF & _CPD_OFF & _CP_OFF & _MCLRE_OFF & _PWRTE_ON & _WDT_OFF
& _HS_OSC
        __CONFIG   _CONFIG2, _WRT_OFF & _BOR21V

   cblock 0x20
Delay1        ; Assign an address to label Delay1
Delay2

COUNTR
BUFFER
DATABYTE
DLYCNT
Display       ; define a variable to hold the diplay
    endc

   org 0
Start:

    bsf    STATUS,RP0    ; select Register Bank 1
```

```
        movlw   0xFF
        movwf   TRISA       ; Make PortA all input
        clrf    TRISD       ; Make PortD all output
        clrf    TRISC
        movlw   0x00        ; Left Justified, Vdd-Vss referenced
        movwf   ADCON1
        bsf     STATUS,RP1  ; select Register Bank 3
        movlw   0xFF        ; we want all Port A pins Analog
        movwf   ANSEL
        bcf     STATUS,RP0  ; back to Register Bank 0
        bcf     STATUS,RP1

        movlw   0x41
        movwf   ADCON0      ; configure A2D for Fosc/8, Channel 0 (RA0), and turn on the
A2D module

MainLoop:
        nop                 ; wait 5uS for A2D amp to settle and capacitor to charge.
        nop                 ; wait 1uS
        nop                 ; wait 1uS
        nop                 ; wait 1uS
        nop                 ; wait 1uS
        bsf     ADCON0,GO_DONE ; start conversion
        btfss   ADCON0,GO_DONE ; this bit will change to zero when the conversion is
complete
        goto    $-1

        movf    ADRESH,w    ; Copy the display to the LEDs
        movwf   PORTD

        movf    ADRESH,w
;       movlw 'A'
        call CONVERT
        call TXD_DATA
        goto    MainLoop

TXD_DATA                    ;Transmit the data to the EVAL232R bridge
    BSF PORTC,6
NEXT CALL DELAY
    RRF BUFFER,1
    BTFSC STATUS,0
    BSF PORTC,6

    BTFSS STATUS,0
    BCF PORTC,6
    DECFSZ COUNTR,1
    GOTO NEXT
    CALL DELAY
    BCF PORTC,6
    CALL DELAY
    CALL DELAY
    RETLW 0

CONVERT                     ;invert data bits to satisfy RS232 protocol
    XORLW 0FFH
    MOVWF BUFFER
```

```
       MOVLW 8
       MOVWF COUNTR
       RETLW 0


DELAY                    ; 104 microsecond delay for 9600 bit
    MOVLW .171
    MOVWF DLYCNT
REDX DECFSZ DLYCNT,1
    GOTO REDX
    NOP
    RETLW 0
end
```

The Object inspector for the Virtual gauge

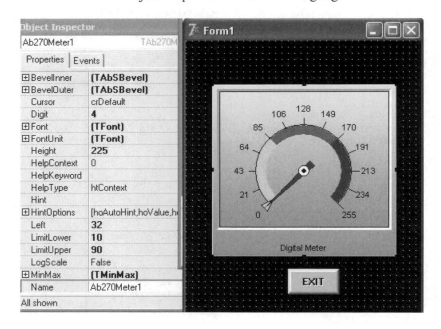

The Delphi project for the gauge can be found in sub folder POTDLL. The Abakus virtual component Ab270Meter is used. Below shows the listing for POTDLLX:

```pascal
unit potdll;

interface

uses
  Windows, Messages, SysUtils, Classes, Graphics, Controls, Forms, Dialogs,
  _GClass, AbRMeter,MMSystem, StdCtrls, AbCBitBt, ExtCtrls;

type
  TForm1 = class(TForm)
    Ab270Meter1: TAb270Meter;
    Timer1: TTimer;
    AbColBitBtn1: TAbColBitBtn;

    procedure FormCreate(Sender: TObject);
    procedure Timer1Timer(Sender: TObject);
    procedure AbColBitBtn1Click(Sender: TObject);
  private
    { Private declarations }
  public
    { Public declarations }
  end;

var
  Form1: TForm1;
  txt:text;
    a:real;
    n:integer;
    data:byte;
    x:word;
    s:string;
    timestr:string;
  FT_Current_Baud:Dword;
   I:Integer;
   B :Byte;
      DevicePresent : Boolean;
  Selected_Device_Serial_Number : String;
  Selected_Device_Description : String;

const

    FT_DLL_NAME='ftd2xx.dll';
    FT_Out_Buffer_Size=1;
implementation
uses D2XXUnit;


{$R *.DFM}



procedure TForm1.FormCreate(Sender: TObject);
begin

Ab270Meter1.digit:=0;
```

```
Timer1.Interval:=1;
Timer1.Enabled:=True;
  Open_USB_Device;

  FT_Enable_Error_Report := false; // Disable Error Reporting



FT_Current_Baud:=FT_BAUD_9600;
FT_Current_DataBits:=FT_DATA_BITS_8;
FT_Current_StopBits:=FT_STOP_BITS_1;
FT_Current_Parity:=FT_PARITY_NONE;
I:= Set_USB_Device_BaudRate;
Set_USB_Device_RTS  ;
Set_USB_Device_DTR  ;

 sndPlaySound('C:\system.wav',
   SND_NODEFAULT Or SND_ASYNC);

AssignFile(txt,'C:\DATAPOT\pot.txt');

Rewrite(txt);
end;

procedure TForm1.Timer1Timer(Sender: TObject);
begin

Timer1.Enabled:=False;
Caption:=TimeToStr(Time)   ;
I:=Purge_USB_Device_In;

n:=0;


 I := Read_USB_Device_Buffer(1);
   data:=FT_In_Buffer[n];
asm
mov al,data
mov x,ax
end;
a:=x;

Ab270Meter1.digit:=data*4;
beep;

timestr:=TimeToStr(Time);

 Str(a,s);
writeln(txt,a:3:1);

Str(a:2:1,s);
Caption:=TimeToStr(Time);

Canvas.TextOut(20,45,s);
```

```
unit led1;
interface
uses
  Windows, Messages, SysUtils, Classes, Graphics, Controls, Forms, Dialogs,
  _Gclass, AbLED;
type
  Tform1 = class(Tform)
    AbLED1: TAbLED;
    AbLED2: TAbLED;
    procedure FormCreate(Sender: Tobject);
  private
    { Private declarations }
  public
    { Public declarations }
  end;
var
  Form1: Tform1;
implementation

{$R *.DFM}
procedure Tform1.FormCreate(Sender: Tobject);
begin
AbLED1.LED.Coloron:=clred;
AbLED1.LED.Coloroff:=clyellow;
AbLED2.LED.Coloroff:=clgreen;
AbLED2.LED.Coloron:=clblue;
AbLED1.LED.Height:=20;
AbLED2.LED.Height:=5;
end;
end.
```

Programming the virtual LED's.

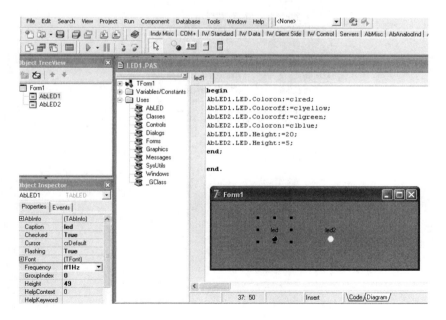

2.4 Delphi Tcolourbutton

The Tcolorbutton adds three new properties to the standard Delphi Tbutton and It can be downloaded from the http://delphi.about.com website and installed as a single new component.

The hovercolor specifies the colour used to paint the buttons background when the mouse pointer hovers over the button. The Forecolor specifies the colour of the button text and the Backcolor specifies the background colour of the button.

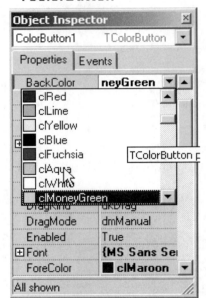

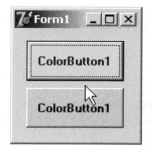 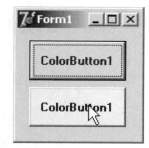

3 VCP USB communications

The next step to achieve virtual instrumentation is to provide a USB communications channel enabling the flow of digital data to and from the virtual components. FTDI's and Prolific RS232/USB bridges are ideally suited to provide a 'virtual communications port' or VCP . These devices are seen by the PC as a standard RS232 COM port although they connect to the USB port. Serial RS232 data can now be sent and received by an external device.

In keeping with all USB interfaces, these RS232/USB modules are no exception and require software drivers in order to function. This chapter shows how to install these drivers and detail the Case Study of a virtual thermometer. Scientific Instruments 'Portcontroller' Active X in conjunction with Delphi gives full control over the serial COM port with just a few lines of code.

3.1 Installing FTDI's VCP drivers

FTDI's EVAL232R shown below is a RS232/USB bridge with a standard USB and 9 pin D-type RS232 connector. The RS232 connector functions just like any RS232 port with TXD and RXD data lines.

EVAL232R FT232RL USB to RS232 Module

Connecting this module via a standard USB cable to the PC's port will invoke the following sequence of operations:

Click on 'No,not this time'.

Click on 'Install from a list or specific location (Advanced)'

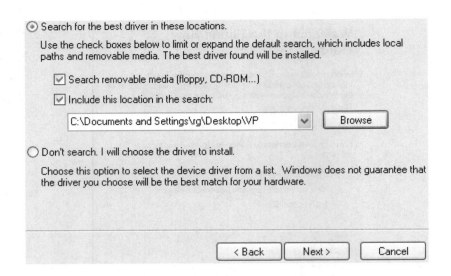

Click on 'Browse' and select the folder which contains the VCP driver.

Click on 'Finish'

The following message box will appear: Note that this time a USB Serial Port is found. The same menu options are presented as before.

This wizard helps you install software for:

USB Serial Port

 If your hardware came with an installation CD or floppy disk, insert it now.

What do you want the wizard to do?

○ Install the software automatically (Recommended)
◉ Install from a list or specific location (Advanced)

Click Next to continue.

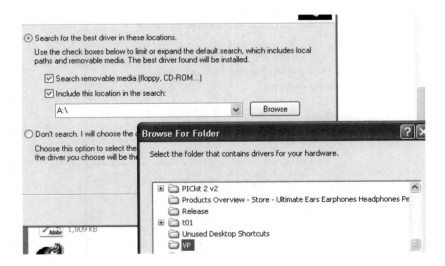

Completing the Found New Hardware Wizard

The wizard has finished installing the software for:

 USB Serial Port

Click Finish to close the wizard.

Now we have to locate which serial COM port has been assigned to the RS232/USB bridge. Select 'System' in Control Panel in Windows and click on 'Hardware', as below:

Click on 'Device Manager'

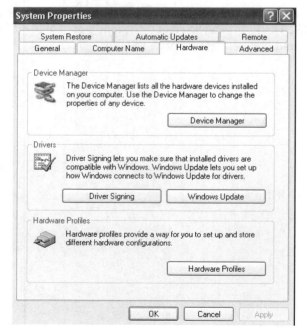

A list of devices is shown and it can be seen that serial port COM5 has been assigned to the USB bridge.

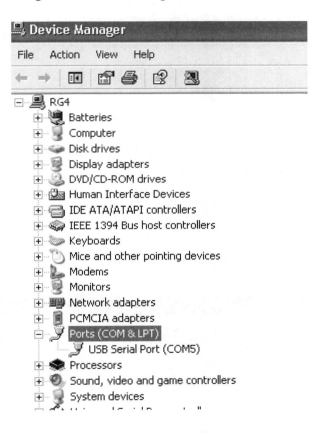

Clicking on 'USB Serial Port (COM5) ' and then on 'Port Settings' and 'Advanced' allows the user to change the port to for example COM port 1 (COM1) or COM2.

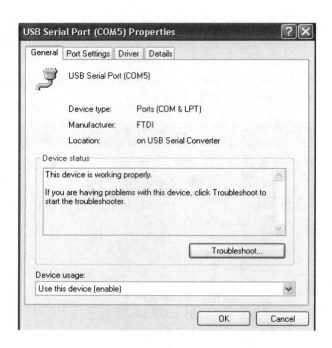

Select COM Port Number e.g. COM1 and click OK

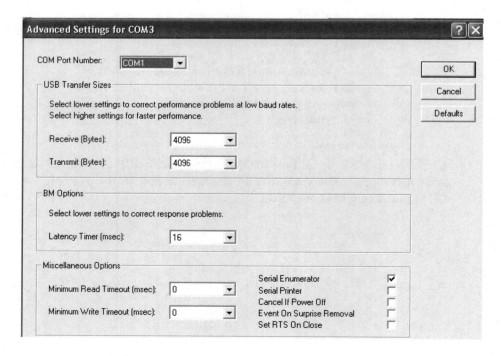

Now the USB bridge can be tested by a simple loopback connecting pins 2 and 3 on the 9 pin D-Type connector, connecting the USB cable from the bridge into the PC's USB port and running a terminal emulation program such as Windows 'Terminal' or 'Superterm'.

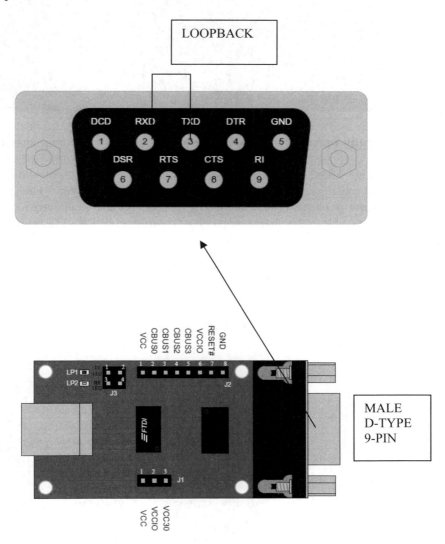

Simply pressing any key on the keyboard will cause that character to be echoed back via the loopback on the RS232 connector of the bridge. There is no need to set the baud rate, since the outgoing data and receiving data will always be at the same baud rate. The letters in red above are the characters pressed on the keyboard and the white letters, the character received from the USB/RS232 bridge.

3.2 RS232 connectors

D-Type 9 pin male and female RS232 connectors

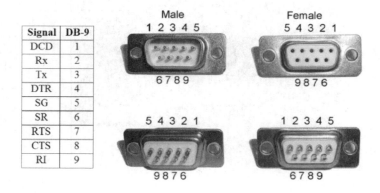

Led indicators can monitor the activity on the TXD and RXD lines of the RS232 port. For this, a converter and an RS232 tester is required as shown below:

9 pin to 25-pin converter   RS232 led Tester

Pin2 txd   Pin3 rxd   Pin7 gnd

A RS232 Tester connected to the EVAL232R module

4 pin and 25 pin gender changers

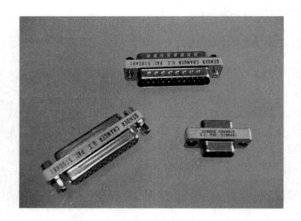

A selection of RS232 connectors and hoods

3.3 Portcontroller Active X

In order to receive data via a VCP USB port an Active X component is required. Scientific Instruments 'Portcontroller' simplifies RS232 serial communications and can be used in Delphi. Its library (see Appendix) contains all the instructions necessary to initialise the COM port and to import and export data via the USB port. Some of the Pascal functions and procedures are listed below:

```
function Get_BaudRate: Integer; safecall;
procedure Set_BaudRate(Value: Integer); safecall;
function Get_DataBits: Smallint; safecall;
procedure Set_DataBits(Value: Smallint); safecall;
function Get_Parity: Smallint; safecall;
procedure Set_Parity(Value: Smallint); safecall;
function Get_StopBits: Smallint; safecall;
procedure Set_StopBits(Value: Smallint); safecall;
function Get_PortName: WideString; safecall;
function Get_Cd: Integer; safecall;
function Get_Cts: Integer; safecall;
function Get_Dsr: Integer; safecall;
function Get_Dtr: Integer; safecall;
procedure Set_Dtr(Value: Integer); safecall;
function Get_DtrDsr: Integer; safecall;
procedure Set_DtrDsr(Value: Integer); safecall;
function Get_RtsCts: Integer; safecall;
procedure Set_RtsCts(Value: Integer); safecall;
function Get_Rts: Integer; safecall;
procedure Set_Rts(Value: Integer); safecall;
```

The trial software download can be found at:
http://www.scientificcomponent.com/
After installing Portcontroller ActiveX, start Delphi and choose Components / Import ActiveX Control from main menu.

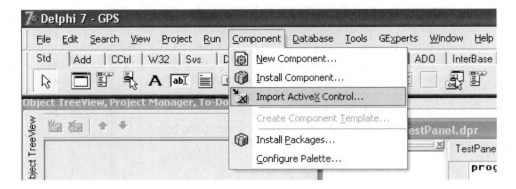

The following dialog appears. Scroll down to the PortController 2.0 entry and select it; Delphi will set the other parameters automatically. Click the Install button.

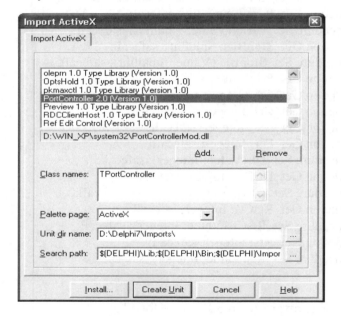

Select Into new package tab page and enter the name of the package file into which you want to install Portcontroller. You may also type description for this package. Click the OK button.

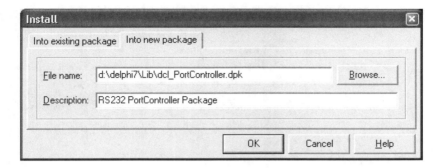

After confirming the message dialogs the Portcontroller package will appear.

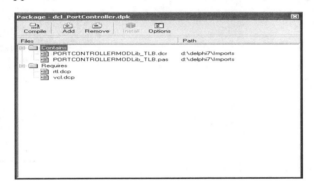

Scroll the Delphi palette to the ActiveX page and confirm that Portcontroller exists on the page.

You may close the Package window. You can create a Portcontroller object by dragging the Portcontroller icon onto designed form. You may now begin coding with Portcontroller. See the Portcontroller Reference at the end of the book for details on how to use Portcontroller properties, methods, and events.

Delphi Program with Active X component installed.

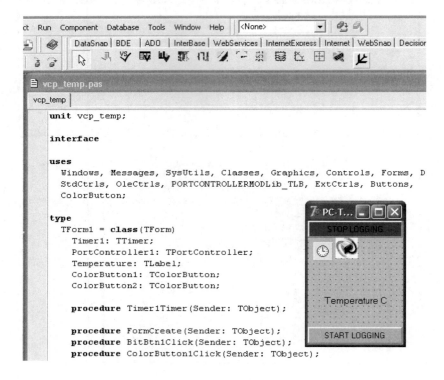

Delphi program VIX uses Portcontroller to import a Byte via a RS232/USB convertor and display on the screen. The Byte, in this case 4EH, can represent an ASCII text character (N) or data. For virtual instrumentation the byte represents a real value –for example, for a 5 volt voltage range, an ADC will produce a byte between 0 and FFh (255).We can then scale this by dividing by 51 so that the displayed result will show a value between 0 and 5 volts i.e. 78 decimal /51 = 1.53.

DATA and TEXT representation of a Byte

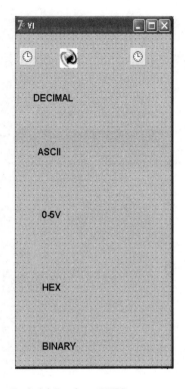 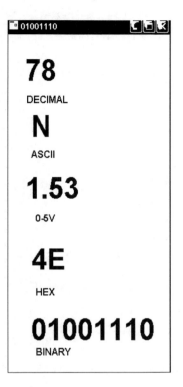

Delphi Project VIX
unit vi;

interface

uses
  Windows, Messages, SysUtils, Classes, Graphics, Controls, Forms, Dialogs,
  StdCtrls, OleCtrls, PORTCONTROLLERMODLib_TLB, ExtCtrls, Buttons,
  ColorButton, ActnList;

type
  Tform1 = class(Tform)
    Timer1: Ttimer;
    PortController1: TportController;
    Label1: Tlabel;
    Label2: Tlabel;
    Label3: Tlabel;
    Label4: Tlabel;
    Label5: Tlabel;
    Timer2: Ttimer;
    Label6: Tlabel;

```
    procedure Timer1Timer(Sender: Tobject);

    procedure FormCreate(Sender: Tobject);

    procedure Timer2Timer(Sender: Tobject);

  private
    { Private declarations }
  public
    { Public declarations }
  end;



var
  Form1: Tform1;

      s:string;

numberbytesread:Integer;
data:string[1];
x:byte;
a:real;

    Buf: array[0..SizeOf(Extended) * 2] of Char;

implementation

{$R *.DFM}


procedure Tform1.Timer1Timer(Sender: Tobject);

begin
  Timer1.Enabled:=False;
PortController1.ClearRQ;
data:=PortController1.Read(1, 0,numberbytesread);
asm
mov ax,data[1];
mov x,al
end;


a:=x;

Canvas.TextOut(30,30,'            ');
Canvas.TextOut(30,120,'           ');
Canvas.TextOut(30,220,'           ');

Canvas.TextOut(40,120,data);    {ASCII}
Str(a:1:0,s);
Canvas.TextOut(30,30,s);        {DECIMAL}
a:=a/51;
Str(a:2:2,s);                   {SCALED}
Canvas.TextOut(30,220,s);
```

```
   BinToHex(@data, Buf, 2);        {HEX}

   Label3.Caption := Format(' %s', [Buf+2]);
   Timer1.Enabled:=True;
   end;

procedure Tform1.FormCreate(Sender: Tobject);
begin
Timer1.Enabled:=True;
 Timer1.Interval:=1;
  PortController1.Open('COM3','9600,8,N,1');

PortController1.RtsCts:=Ord(true);
  PortController1.DtrDsr :=Ord(true);
end;


Function HexToBin(Hexadecimal: string): string;
const
  BCD: array [0..15] of string =
    ('0000', '0001', '0010', '0011', '0100', '0101', '0110', '0111',
    '1000', '1001', '1010', '1011', '1100', '1101', '1110', '1111');
var
 i: integer;

begin
 for i := Length(Hexadecimal) downto 1 do

   Result := BCD[StrToInt('$' + Hexadecimal[i])] + Result;
end;

procedure Tform1.Timer2Timer(Sender: Tobject);
begin
Caption:=(HexToBin(Buf+2));     {BINARY}
Canvas.TextOut(40,450,Caption);
end;

end.
```

3.4 Case study VCP DS75 virtual digital and analogue thermometer

Digital Thermometer readout

 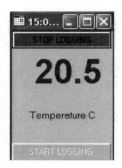

Our first virtual instrument requires three elements – a USB/RS232 bridge, a PIC controlled temperature sensor and a Delphi virtual PC digital thermometer. The sensor used is a DS75 digital thermometer and thermostat which provides 9,10,11 or 12-bit digital temperature readings over a –55 degree C to +125 degree C. Communications between it and a PIC12F629 is achieved via a simple 2-wire serial interface. Three address pins allow up to eight DS75 devices to operate on the same 2-wire bus, which greatly simplifies distributed temperature sensing applications.

The DS75

PIN ASSIGNMENT

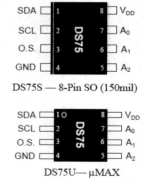

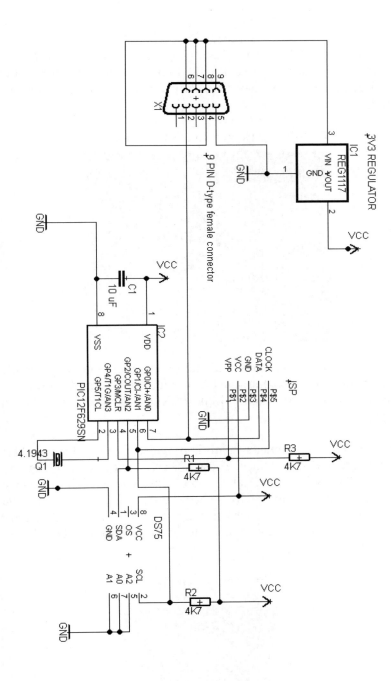

Schematic diagram Digital Thermometer

In this cicuit, FTDI's EVAL232 bridge is used configured as a VCP device and connected via a RS232 D-type connector. Indeed any RS232/USB convertor may be used including the Prolific and the Belkin bridge.

Prolific and Belkin RS232/USB converters

The Thermometer displays a digital temperature and also logs the temperature at 1 second intervals into file 'Temperature.TXT' which then can be imported into EXCEL and displayed as a graphical trend.

Power to the Thermometer board is provided by the RTS and DTR lines of the RS232 interface. These lines are connected together and fed to a 3V3 voltage regulator. This helps sync with the PC since the board only becomes active when the PC's software enables the RTS/DTR hardware handshake lines. Finally an ISP (in circuit programming) connector is used to connect to a PICKIT 2 PIC programmer.

The Thermometer board

The DS75 device stores the temperature in a 2 byte internal register. The high byte holds the temperature value of + 0 to 125 degrees, the low byte holds fractional temperature values. Below zero degrees C the value is stored as a 2's compliment number. The table below shows the register values for several different temperatures.

Figure 3. TEMPERATURE, T_H, and T_L REGISTER FORMAT

	bit 15	bit 14	bit 13	bit 12	bit 11	bit 10	bit 9	bit 8
MS Byte	S	2^6	2^5	2^4	2^3	2^2	2^1	2^0

	bit 7	bit 6	bit 5	bit 4	bit 3	bit 2	bit 1	bit 0
LS Byte	2^{-1}	2^{-2}	2^{-3}	2^{-4}	0	0	0	0

Table 3. 12-BIT RESOLUTION TEMPERATURE/DATA RELATIONSHIP

TEMPERATURE (°C)	DIGITAL OUTPUT (BINARY)	DIGITAL OUTPUT (HEX)
+125	0111 1101 0000 0000	7D00h
+25.0625	0001 1001 0001 0000	1910h
+10.125	0000 1010 0010 0000	0A20h
+0.5	0000 0000 1000 0000	0080h
0	0000 0000 0000 0000	0000h
-0.5	1111 1111 1000 0000	FF80h
-10.125	1111 0101 1110 0000	F5E0h
-25.0625	1110 0110 1111 0000	E6F0h
-55	1100 1001 0000 0000	C900h

For example a +ve temperature of 21.75 degrees would be stored as:

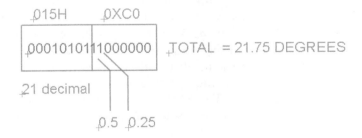

For a temperature below zero, (bit 15 of the data will be a logic 1) the registers for a temperature of – 10.125 degrees shown below, a 2's complement operation has to be performed on the data to extract the temperature value. It was decided to do this not in the PIC software but in the Delphi program.

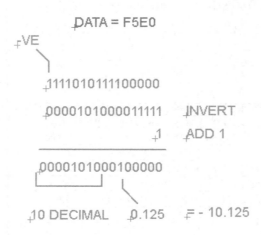

The PIC software- sensor33-2400.ASM and the HEX code for the digital thermometer can be downloaded from the ELEKTOR.COM Website and can be found in Folder vcp_prolific_ftdi_temp.

The Delphi Project (vcp_ftdi_temp) with file vcp_temp.pas runs on a PC and is a software program that replicates the image and functionality of a digital thermometer. It does not use any visual instrument components and only uses the colorbutton component as buttons. Once the thermometer board is connected to the PC via a standard USB lead, running the program and clicking on the start button starts operation. To test the thermometer, freezer spray was applied to the sensor and then a soldering iron tip was applied for a few seconds. The results are shown below: The X axis calibrated in degrees C, The Y axis number of samples at 1 sample/sec.

Freezer and heat test for the thermometer.

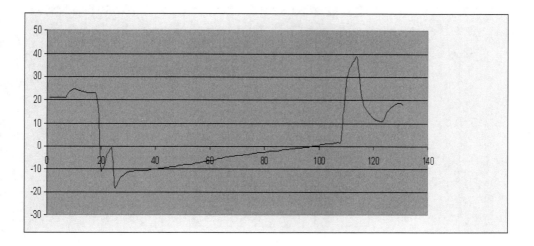

To create the graph, open file temperature .TXT in EXCEL, click on the A column which will automatically highlight all the data values in the column and then, click on the chart symbol and select the type of graph. The X axis is calibrated in degrees C and the Y axis in number of samples.

Importing logged temperature data into EXCEL

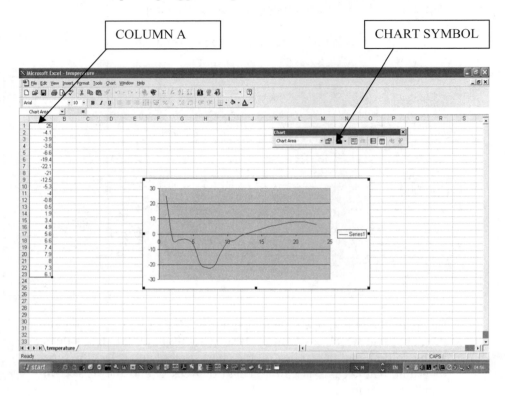

The Delphi Project 'vcp_ftdi_temp' for the virtual digital thermometer in Folder vcp_prolific_ftdi_temp.:

(a)

```
unit vcp_temp;

interface

uses
  Windows, Messages, SysUtils,
Classes, Graphics, Controls, Forms,
Dialogs,
  StdCtrls, OleCtrls,
PORTCONTROLLERMODLib_TLB,
ExtCtrls, Buttons,
  ColorButton;

type
  TForm1 = class(TForm)
    Timer1: TTimer;
    PortController1: TPortController;
    Temperature: TLabel;
    ColorButton1: TColorButton;
    ColorButton2: TColorButton;

    procedure Timer1Timer(Sender:
TObject);

    procedure FormCreate(Sender:
TObject);
    procedure BitBtn1Click(Sender:
TObject);
    procedure
ColorButton1Click(Sender: TObject);
    procedure
ColorButton2Click(Sender: TObject);
  private
    { Private declarations }
  public
    { Public declarations }
  end;



var
  Form1: TForm1;

    s:string;
```

(b)

```
numberbytesread:Integer;
data:string[1];
x:word;
a:real;
msb:byte;
lsb:byte;
n:byte;
samples:integer;
txt:text;
flag:boolean;
timestr:string;
implementation

{$R *.DFM}


procedure
TForm1.Timer1Timer(Sender:
TObject);
LABEL OUT;
begin
Timer1.Enabled:=False;
flag:=true;
n:=0;
data:=PortController1.Read(1,
0,numberbytesread);
{data:='!'; }
asm
push ax
mov ax,data[1];
mov msb,al;
pop ax
end;
data:=PortController1.Read(1,
0,numberbytesread);
{data:='!';
```

(c)
asm

push ax
push bx

mov ax,data[1];
mov lsb,al;
mov ah,msb;
CMP AH,0;
JNs @OUT
XOR AX,0ffffh
ADD AX,1
mov bl,1
mov n,bl
@OUT:
shr ax,1;
shr ax,1;
shr ax,1;
shr ax,1;
shr ax,1;
mov x,ax;

pop bx
pop ax
end;

a:=x;
a:=a/8;
timestr:=TimeToStr(Time);
{write(txt,' ',timestr,' '); }
if n=1 then
write(txt,'-');
writeln(txt,a:2:1);
Str(a:2:1,s);
Caption:=TimeToStr(Time);
Canvas.TextOut(10,30,'   ');
if n=1 then
Canvas.TextOut(10,29,'-');
Canvas.TextOut(30,30,s);
flag:=false;
 Timer1.Enabled:=True;
end;

(d)

procedure TForm1.FormCreate(Sender: TObject);
begin
Timer1.Enabled:=False;
 Timer1.Interval:=10;
PortController1.RtsCts:=Ord(true);
PortController1.DtrDsr:=Ord(true);
AssignFile(txt,'C:\remote.txt');
Rewrite(txt);
flag:=false;
end;

procedure
TForm1.BitBtn1Click(Sender: TObject);
begin
Timer1.Enabled:=False;
end;

procedure
TForm1.ColorButton1Click(Sender: TObject);
begin
{while flag= false do }
ColorButton1.Enabled := False;
if flag=false then
Timer1.Enabled:=False;
PortController1.Close;
CloseFile(txt);
close;
end;

procedure
TForm1.ColorButton2Click(Sender: TObject);
begin

PortController1.Open('COM2','2400, 8,N,1');
Timer1.Enabled:=tRUE;
ColorButton2.Enabled := False;
end;
end.

7.2 The Compass PC Software.

Delphi Project file COMPASSX2.PAS can be downloaded from Folder 'COMPASS', subfolder 'COMPASSX' The virtual Compass component (Abcompass) in environment options AbAnalogueIng can be downloaded from the 'ABACUS' website @ http://www.abaecker.biz/abkdownload.html.

The Delphi Compass Form

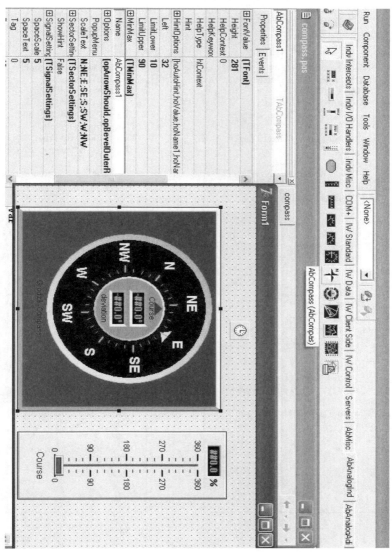

(a)
PROGRAM COMPASSX2.PAS

unit compass2;

interface

uses
  Windows, Messages, SysUtils, Classes, Graphics, Controls, Forms, Dialogs,
  _GClass, AbCompas, ExtCtrls, AbHSlide, AbVSlide;

type
  TForm1 = class(TForm)
    AbCompass1: TAbCompass;
    Timer1: TTimer;
    AbVSlider1: TAbVSlider;
    procedure FormCreate(Sender: TObject);
    procedure Timer1Timer(Sender: TObject);
  private
    { Private declarations }
  public
    { Public declarations }
  end;

var
  Form1: TForm1;
txt:text;
  a:real;
  n:integer;
  data:byte;
  x:word;
  s:string;
timestr:string;
  FT_Current_Baud:Dword;
  I:Integer;
  B :Byte;
    DevicePresent : Boolean;
  Selected_Device_Serial_Number : String;
  Selected_Device_Description : String;

(b)
const
    FT_DLL_NAME='ftd2xx.dll';
    FT_Out_Buffer_Size=1;
implementation
uses D2XXUnit;

{$R *.DFM}

procedure TForm1.FormCreate(Sender: TObject);
begin
AbCompass1.ValueShould :=0;
AbvSlider1.Value:=0;
AbCompass1.digit:=0;

Canvas.Pen.Color:=ClLime;
Timer1.Interval:=250;
Timer1.Enabled:=True;
  Open_USB_Device;

 FT_Enable_Error_Report := false; // Di250sable Error Reporting

FT_Current_Baud:=FT_BAUD_9600;
FT_Current_DataBits:=FT_DATA_BITS_8;
FT_Current_StopBits:=FT_STOP_BITS_1;
FT_Current_Parity:=FT_PARITY_NONE;
 I:= Set_USB_Device_BaudRate;

end;

procedure TForm1.Timer1Timer(Sender: TObject);
begin
Timer1.Enabled:=False;
Caption:=TimeToStr(Time)   ;
I:=Purge_USB_Device_In;

AbCompass1.ValueShould := AbvSlider1.Value;
n:=0;

©

```
 I := Read_USB_Device_Buffer(1);
   data:=FT_In_Buffer[n];
asm
mov al,data
mov x,al
end;
if AbCompass1.deviation > 2  then
begin

beep;
end;
if  AbCompass1.deviation < -2  then
begin
beep;

end;


a:=x*1.41176;

AbCompass1.digit:=data*4;


timestr:=TimeToStr(Time);

 Str(a,s);


Str(a:2:0,s);
Caption:=TimeToStr(Time);
Canvas.TextOut(440,60,'      ');
Canvas.TextOut(440,60,s);

Timer1.Enabled:=True;
end;

end.
```

7.3 Operating the compass

Rotating the hardware compass will cause the compass rose on the PC's screen to rotate in sympathy and will show the magnetic heading. If the course deviation i.e. the difference between the compass heading and the target course, is greater than +- 2 degrees an audible beep is heard. In compass A below, both the compass heading and the target course heading are identical at 282 degrees- the deviation shows a 0.5 degree error and no beep is issued.

Compass   A: :

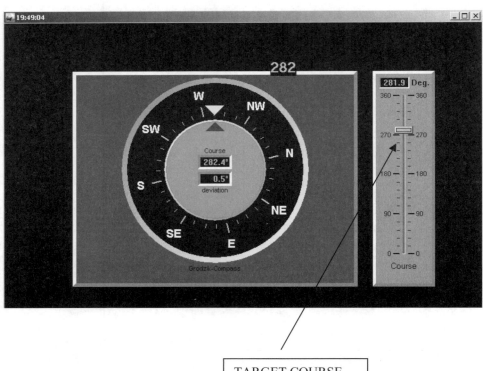

TARGET COURSE

In compass B, the target course heading is set at 177 degrees, with a heading of 268 degrees, producing an error of 90 degrees and a corresponding beep.

Compass   B:

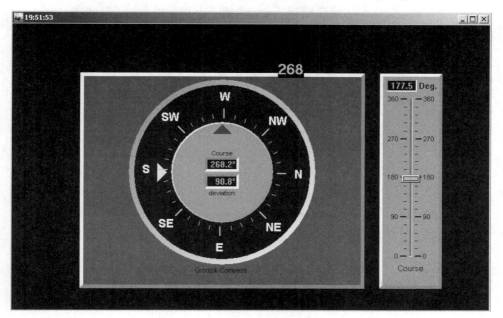

7.4 Testing the Compass

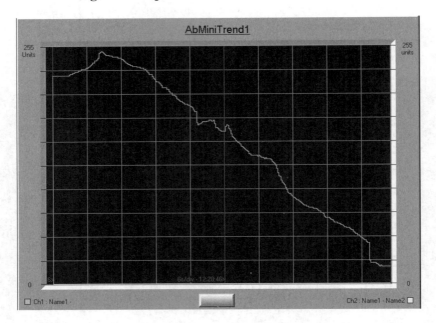

Program 'Compasstest' in folder 'Compass' subfolder 'compasstest' is used to test the compass. Rotating the compass slowly through 360 degrees will produce the trend graphics shown below. Note that this is not a straight line because of the inevitable jerkiness of the hand when moving the compass. The X axis has a maximum value of 255 corresponding to 0/360 degrees.

8 Case study FFT audio frequency analyser

The DSP PIC30F6012A

This development board is based on Microchip's dsPIC30F6012A 30 Mips 64 pin TQFP processor which contains extensive Digital Signal Process (DSP) functions with a high performance 16-bit microcontroller (MCU) architecture. Although DSP and FFT functions require a high level of maths, example programs are given which includes an FFT procedure written by Microsoft, so that the user can explore the use of this DSP chip with little effort. It includes the use of the 12 bit ADC and also the UART which permits the results of the FFT (Peak Frequency and Bin number) to be displayed on a PC and displayed as a peak frequency spectrum.

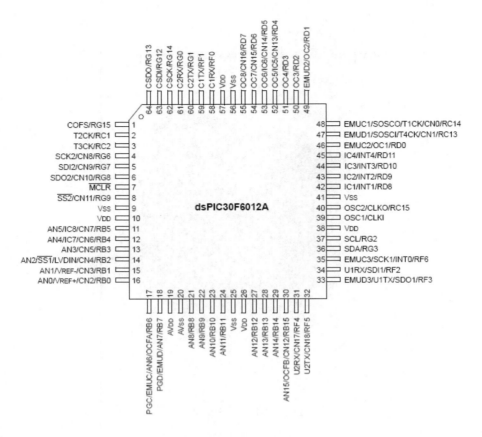

Pin out

Besides the dsPIC30F devices having a DSP engine, these 16-bit embedded controllers may also be used as general purpose processors, having a plethora of user ports. : RB0-RB15, RG0-RG15, RD0-RD11 and RF0-RF6. With a crystal frequency of 7.12 Mhz and an internal PLL of up to x 16, this chip can achieve 30 Mips. At these frequencies, the chip gets hot and consumes as much as 250 mA, but a much lower 3mA consumption is possible with a 3V3 supply voltage and using the LPRC 512 Khz internal oscillator.

This 'no frills' board could not be simpler. A surface mount dsPIC30F6012a processor, a couple of LED's: (LED0 the ubiquitous flashing (blue) status indicator, and LED1 a red power indicator), FTDI's MM232R USB/RS232 bridge for PC communications (VCP) @ 115200 baud, and a reset switch (SW1). In addition an in-circuit header to enable the programming of the board using the PICkit 2 programmer. Finally, a simple phono socket to connect the signal source (audio frequencies) to the ADC pin AN3. Power is taken from the USB port and provides a stable 5 volt d.c. source.

MM232R MINI USB-SERIAL MODULE

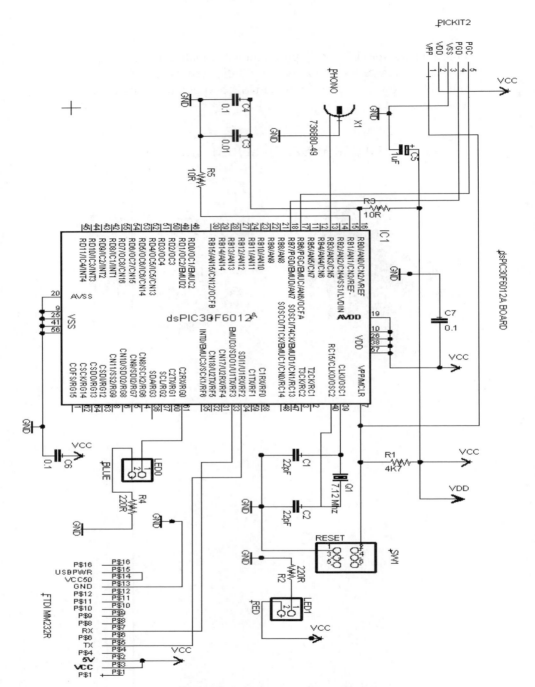

schematic dsPIC30F6012a development Board

Later we shall show how a virtual representation of the audio frequency spectrum can be obtained.(shown below). A sine wave of 906 Hz was applied to the phono input of the FFT board, the resulting bin numbers and the relative amplitudes are shown. Note that with a sampling frequency of 8000 Hz, and a sampling window of 255 bytes, each Bin number represents a frequency multiply of 31.25Hz.

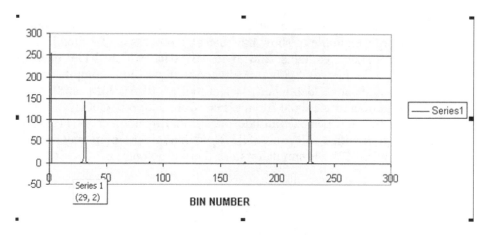

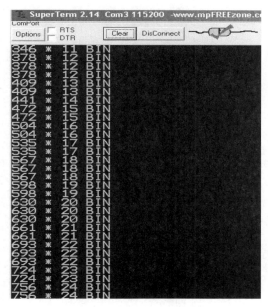

A simple terminal emulator 'Super Term' is also used to display the input frequency and bin number. Above the screen picture shows that a frequency generator shows the resultant frequencies 346 HZ to 756Hz and their associated bin numbers (11 to 24).

8.1 FFT SIMULATOR

Mplab Project 'Simul.mcw' in Folder 'DSP' is based on Microchip's example to show how the FFT process is executed. File 'square250Hz.c' contains 256 bytes of data which have been sampled at a rate of 8000 times/second. If we now compile the program and select 'debugger', 'select tool', MPLAB SIM' and run the program, and view the file registers from address 1C00H, it can be seen that the peak amplitude occurs in Bin number 08 (1C00 being Bin 0) with a value of 03DH. Note that this program may only be run in the MPLAB simulator since there is no output.

Mplab Project Simul

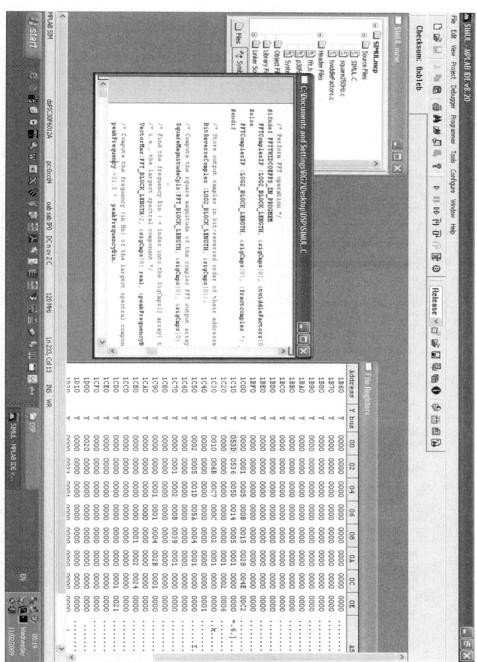

Program SIMUL.C (Folder DSP)

```c
#include "p30F6012a.h"


#include <p30Fxxxx.h>
#include "delay.h"
#include <dsp.h>
#include "fft.h"
#include "stdlib.h"
#include "string.h"
#include "ctype.h"
void convertpeak(void) ;
void FFT(void);

//Functions and Variables with Global Scope:
void ADC_Init(void);
void __attribute__((__interrupt__)) _ADCInterrupt(void);
void convertADC(void) ;


int Data_ADC =0;

int rep;
//Functions:

//ADC_Init() is used to configure A/D to scan and convert 1 input channel AN3
//per interrupt. The A/D is set up for a total sampling rate of 8 KHz


int unsigned short  u;
int unsigned short t;
int unsigned short h;
int unsigned short th;
 char ThisPage[28];
 int   number;

 int str8 ;
 int str9 ;
 int str10 ;
 int str11 ;


 int str13 ;

int unsigned short org;
int samples [256];
int roms;

unsigned int ADResult1 = 0;
unsigned int ADResult2 = 0;

//Functions and Variables with Global Scope:
void ADC_Init(void);
void __attribute__((__interrupt__)) _ADCInterrupt(void);
void ADCResult2Decimal(unsigned int ADRES);
```

```c
// © 2007 Microchip Technology Inc.
/* Device configuration register macros for building the hex file */
_FOSC(CSW_FSCM_OFF & XT_PLL8);      /* XT with 8xPLL oscillator, Failsafe clock off */
_FWDT(WDT_OFF);                     /* Watchdog timer disabled */
_FBORPOR(PBOR_OFF & MCLR_EN);       /* Brown-out reset disabled, MCLR reset enabled */
_FGS(CODE_PROT_OFF);                /* Code protect disabled */


/* Extern definitions */
extern fractcomplex sigCmpx[FFT_BLOCK_LENGTH]         /* Typically, the input signal to an FFT */
__attribute__ ((section (".ydata, data, ymemory"),    /* routine is a complex array containing samples */
aligned (FFT_BLOCK_LENGTH * 2 *2)));                  /* of an input signal. For this example, */
                                                      /* we will provide the input signal in an */
                                                      /* array declared in Y-data space. */
/* Global Definitions */
#ifndef FFTTWIDCOEFFS_IN_PROGMEM
fractcomplex twiddleFactors[FFT_BLOCK_LENGTH/2]       /* Declare Twiddle Factor array in X-space*/
__attribute__ ((section (".xbss, bss, xmemory"), aligned (FFT_BLOCK_LENGTH*2)));
#else
extern const fractcomplex twiddleFactors[FFT_BLOCK_LENGTH/2]   /* Twiddle Factor array in Program memory */
__attribute__ ((space(auto_psv), aligned (FFT_BLOCK_LENGTH*2)));
#endif

int          peakFrequencyBin=0;                /* Declare post-FFT variables to compute the */
int unsigned short peakFrequency ;              /* frequency of the largest spectral component */

int unsigned short peakFrequency2;



#define XTFREQ      7372800         //On-board Crystal frequency
#define PLLMODE     8               //On-chip PLL setting
#define FCY         XTFREQ*PLLMODE/4  //Instruction Cycle Frequency

#define BAUDRATE    115200          //SELECT BAUD RATE HERE  115200
#define BRGVAL      ((FCY/BAUDRATE)/16)-1

int main(void)

{
```

```
    asm("nop");

        ADPCFG = 0xFF;                      //Make analog pins digital
        LATG = 0x0;
        TRISG = 0x0;                        //Configure LED pins as output
    PORTG = 0;
        TMR1 = 0;                           // clear timer 1
        PR1 = 0x7270;                       // interrupt every 250ms
        IFS0bits.T1IF = 0;                  // clr interrupt flag


        IEC0bits.T1IE = 0;                  // set interrupt enable bit


        T1CON = 0x8030;                     // Fosc/4, 1:256 prescale, start TMR1

        TRISF = 0xC;
        U1BRG  = BRGVAL;
        U1MODE = 0x8000;                    // Reset UART to 8-n-1, alt pins, and enable
        U1STA  = 0x0440;                    // Reset status register and enable TX & RX


        IEC0bits.U1TXIE = 0;  //DISABLE UART INTERRUPTS
    IEC0bits.U1RXIE = 0;  //DISABLE UART INTERRUPTS

        _U1RXIF=0;                                                  // Clear UART RX Interrupt Flag


    FFT();


here:
 goto here;

 }

void __attribute__((__interrupt__, no_auto_psv)) _T1Interrupt(void)
{
        IFS0bits.T1IF = 0;    // clear interrupt flag



}


// © 2007 Microchip Technology Inc.
void FFT(void)
{
        int i = 0;
        fractional *p_real = &sigCmpx[0].real ;
        fractcomplex *p_cmpx = &sigCmpx[0] ;
```

```c
#ifndef FFTTWIDCOEFFS_IN_PROGMEM
    /* Generate TwiddleFactor Coefficients */
    TwidFactorInit (LOG2_BLOCK_LENGTH, &twiddleFactors[0], 0);         /* We need to do this only once at start-up */
#endif

    for ( i = 0; i < FFT_BLOCK_LENGTH; i++ )/* The FFT function requires input data */
    {                                       /* to be in the fractional fixed-point range [-0.5, +0.5]*/


        *p_real = *p_real >>1 ;             /* So, we shift all data samples by 1 bit to the right. */
        *p_real++;                          /* Should you desire to optimize this process, perform */
    }                                       /* data scaling when first obtaining the time samples */
                                            /* Or within the BitReverseComplex function source code */

    p_real = &sigCmpx[(FFT_BLOCK_LENGTH/2)-1].real ;    /* Set up pointers to convert real array */
    p_cmpx = &sigCmpx[FFT_BLOCK_LENGTH-1] ; /* to a complex array. The input array initially has all */
                                            /* the real input samples followed by a series of zeros */

    for ( i = FFT_BLOCK_LENGTH; i > 0; i-- ) /* Convert the Real input sample array */
    {                                       /* to a Complex input sample array */
        (*p_cmpx).real = (*p_real--);       /* We will simpy zero out the imaginary */
        (*p_cmpx--).imag = 0x0000;          /* part of each data sample */
    }

    /* Perform FFT operation */
#ifndef FFTTWIDCOEFFS_IN_PROGMEM
    FFTComplexIP (LOG2_BLOCK_LENGTH, &sigCmpx[0], &twiddleFactors[0], COEFFS_IN_DATA);
#else
    FFTComplexIP (LOG2_BLOCK_LENGTH, &sigCmpx[0], (fractcomplex *) __builtin_psvoffset(&twiddleFactors[0]), (int) __builtin_psvpage(&twiddleFactors[0]));
#endif

    /* Store output samples in bit-reversed order of their addresses */
    BitReverseComplex (LOG2_BLOCK_LENGTH, &sigCmpx[0]);

    /* Compute the square magnitude of the complex FFT output array so we have a Real output vetor */
    SquareMagnitudeCplx(FFT_BLOCK_LENGTH, &sigCmpx[0], &sigCmpx[0].real);
```

```
        /* Find the frequency Bin ( = index into the SigCmpx[] array) that has the largest
energy*/
           /* i.e., the largest spectral component */
           VectorMax(FFT_BLOCK_LENGTH/2, &sigCmpx[0].real, &peakFrequencyBin);

           /* Compute the frequency (in Hz) of the largest spectral component */
           peakFrequency =31.5 *  peakFrequencyBin;
 /*  samling frequency/no. samples 29.35    */

  peakFrequency2=  peakFrequency   ;


}

 void delay(unsigned t)

{

T1CON=0x8000;

  while (t--)

  {TMR1=0;
   while (TMR1<1600);
  }

  }
```

8.2   FFT with ADC and serial output

Mplab project 'DSPX' utilises the PIC onboard ADC to sample an input audio signal and then to provide a visual output by streaming the FFT data via the FTDI MM232R RS232/USB adapter to a PC terminal emulator.

Since the development board has no input signal conditioning circuitry to the ADC AN3 pin, a suitable signal source has to be selected to provide a signal. Maplins function generator Type N42FL was chosen since it provide sine, square, triangle and sawtooth waveforms at a frequency range of 100 Hz to 20 Khz. A small modification needs to be made since the phono output only provides a fixed 5 volt output.To make the modification, open circuit the phone signal line and connect the X2 'speaker' output to the phono socket. The amplitude of the signal can then be varied by the R19 'volume' potentiometer.

Function Generator Kit
MAP 432 - N42FL

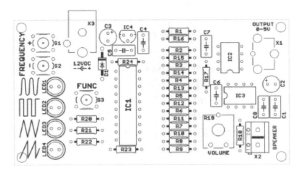

- This microcontroller based function generator is capable of producing square, sine, sawtooth or triangle waves. The wave shape and frequency are adjustable via on board push buttons. A 0-5V output is available as well as an audio output that can drive a small speaker, up to 1W.

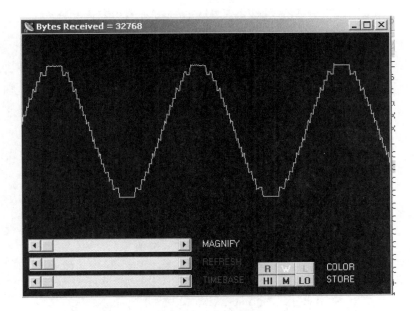

Typical output from the Maplin's function generator.

The next stage is to download the HEX file DSPX.HEX from Folder 'DSP' from the ELEKTOR website and program the dsPIC30F6012A processor using the PICkit 2 programmer. Note that this programmer will not support type dsPIC30F6012. An alternative programmer may be used if the device is supported.

Connect in a phono lead between the function generator and the development board. Also connect a USB cable between the MM232R bridge and the PC and run any terminal emulation program. The software runs in a continuous loop, the blue LED should flash @ 500 milliseconds intervals and increasing the output frequency of the signal generator should produce the following on the screen.

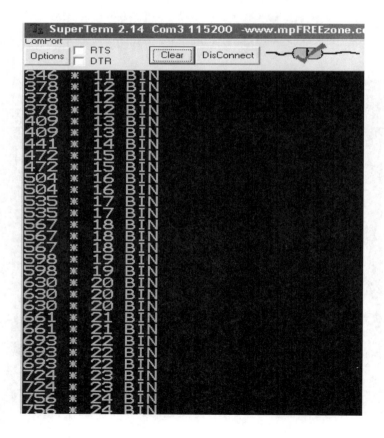

The Software and DSP FFT.
The program was written in the MPLAB environment, with The C30 compiler for dsPIC devices having been previously installed. A free student version is available from Microchip's website. Project File 'DSPX.mcw' can be found in Folder 'DSP'

The MPLAB Project window shows 3 source files:

DSP.c (This is the main program)
Square-4Khz.c (imported example of a square wave)
TwiddleFactors.c (part of the FFT process)

The original program was based on a Microchip example which performed a FFT on the resident square wave data, so that no ADC is involved. This time, however, the 256 bytes for the square 4Khz.c table are overwritten by the ADC results. The 12 bit ADC will collect 256 samples @ a rate of 8000/second, store them in an array 'ROM' and also write the ADC data to file memory. Once the FFT process is completed, the resulting Peak frequency and Bin number will be transmitted from the board via the MM232R to the terminal emulator on the PC at 115200 baud. What is the Bin number?. With a sampling frequency of 8000 samples/second and 256 samples being collected by the ADC, each bin will represent 8000/256 =31.25 Hz. So for example with a bin number of 34, (34 x 31.25) = 1062.2 Hz iss generated i.e. the fundamental frequency of the input signal. What we have done is used the DSP processing capabilities of the dsPIC30F6012a to perform a FFT transform on a signal to convert from a time to a frequency domain.

Code snippet from 'DSP.C'
//setup module address operation to write ADC data to memory file register buffer at start address 0x1c00

```
asm("mov #0x1c00,w0");
asm ("mov w0,XMODSRT");
 asm("mov #0x1fff,w0");  //1024bytes
asm("mov w0,XMODEND");
asm("mov #0x8001,w0");
asm("mov w0,MODCON");
```

and to write ADC data to file registers:

```
asm("mov ADCBUF0,w0");

asm("mov w0,[w10++]");  //write ADC data to 256 file registers
```

Code snippet from 'DSP.C'
//Initialise the 12 bit ADC

```c
void ADC_Init(void)
{
    ADCHSbits.CH0SA=3;
    ADCON1bits.SSRC = 7;
    ADCON1bits.ASAM = 1;

    //ADCON2 Register
    //Set up A/D for interrupting after1 samples get filled in the buffer
    //Also, disable Channel scanning,REF VOLTS =Vcc.
    //All other bits to their default state
    ADCON2bits.SMPI = 1;

    ADCON1bits.FORM = 3;

    ADCON2bits.CSCNA = 0;
    ADCON2bits.VCFG = 0;

    ADCON3bits.SAMC = 31;
    ADCON3bits.ADCS = 40 ;


    ADPCFGbits.PCFG3 = 0;


    //Clear the A/D interrupt flag bit
    IFS0bits.ADIF = 0;

    //Set the A/D interrupt enable bit
    IEC0bits.ADIE = 1;


    //Turn on the A/D converter
    //This is typically done after configuring other registers
    ADCON1bits.ADON = 1;

}
```

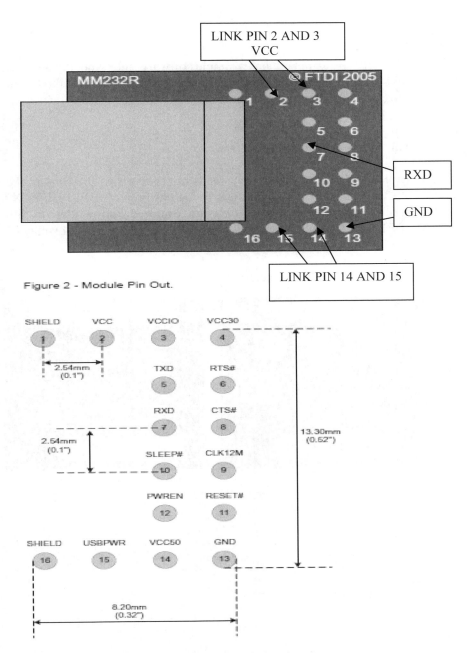

Figure 2 - Module Pin Out.

Figure 3 - MM232R Signal pin out and PCB pad layout.

connections to the MM232R adaptor

The DSP software contains a simple software filter whereby a signal of 440Hz +- 2 Hz will cause the blue LED to stay on and thus form the basis of a simple guitar tuner.

8.3 Virtual spectrum display

Mplab Program 'Delphi.mcw' reads the contents of the Bins in the PIC file register and sends them to Delphi Project 'DELPHIFFTX' where 256 bytes of BIN data are captured and written to file 'FFT.txt'. This file is imported into EXCEL and a virtual display of the frequency spectrum is produced.

The FFT spectrum logger 'DELPHIFFTX'

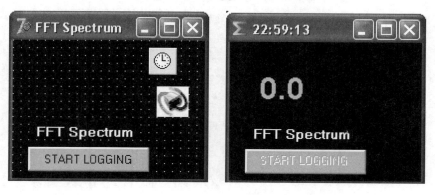

625Hz Sine wave

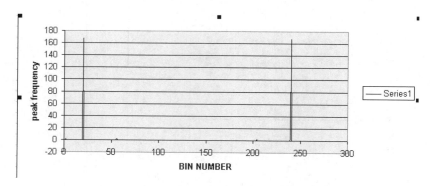

The square wave frequency spectrum shown below has harmonics displayed, though at a much reduced amplitude compared to the fundamental frequency.

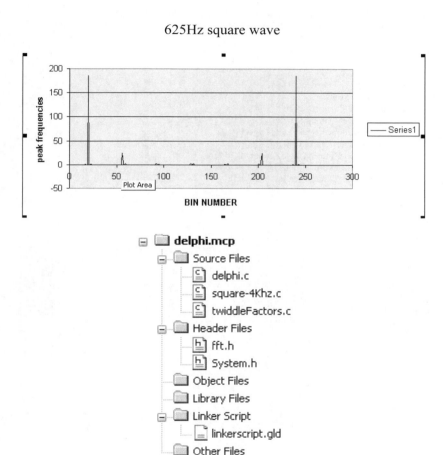

Program DelphiFFTX.PAS

```pascal
unit delphifft;
interface
uses
  Windows, Messages, SysUtils, Classes, Graphics, Controls, Forms, Dialogs,
  StdCtrls, OleCtrls, PORTCONTROLLERMODLib_TLB, ExtCtrls,MMSystem, Buttons,
  ColorButton;
type
  TForm1 = class(TForm)
    Timer1: TTimer;
    PortController1: TPortController;
    Temperature: TLabel;
    ColorButton2: TColorButton;
    procedure Timer1Timer(Sender: TObject);
    procedure FormCreate(Sender: TObject);
    procedure BitBtn1Click(Sender: TObject);
    procedure ColorButton1Click(Sender: TObject);
    procedure ColorButton2Click(Sender: TObject);
    procedure Timer2Timer(Sender: TObject);
  private
    { Private declarations }
  public
    { Public declarations }
  end;

var
  Form1: TForm1;
    s:string;
numberbytesread:Integer;
data:string[1];
x:word;
a:real;
rep:byte;
msb:byte;
lsb:byte;
n:byte;
txt:text;
flag:boolean;
timestr:string;
implementation

{$R *.DFM}


procedure TForm1.Timer1Timer(Sender: TObject);
LABEL OUT;
begin
PortController1.RtsCts:=Ord(true);
PortController1.DtrDsr:=Ord(true);

flag:=true;
Timer1.Enabled:=False;

OUT:
data:=PortController1.Read(1,0,numberbytesread);
```

```
{data:='!'; }
asm
push ax
mov ax,data[1];
mov msb,al;
mov al,msb
mov ah,0
mov x,ax
pop ax
end;

a:=x;
if a=255 then   {Trigger to start data capture}
begin
goto OUT;
end;

if n=1 then
write(txt,'-');
writeln(txt,a:3:1);


Str(a:2:1,s);
Caption:=TimeToStr(Time);
Canvas.TextOut(10,30,'             ');
if n=1 then
Canvas.TextOut(10,29,'-');
Canvas.TextOut(30,30,s);
flag:=false;
Timer1.Enabled:=true;
 rep:=rep+1;
 if rep=255  then
begin
 Timer1.Enabled:=False;
CloseFile(txt);
Timer1.Enabled:=False;
PortController1.Close;
 close;
end;
end;

procedure TForm1.FormCreate(Sender: TObject);

begin
Timer1.Enabled:=False;
Timer1.Interval:=1;

AssignFile(txt,'fft.txt')  ;
Rewrite(txt);
flag:=false;
beep;
end;

procedure TForm1.BitBtn1Click(Sender: TObject);
begin
```

```
Timer1.Enabled:=False;
end;

procedure TForm1.ColorButton1Click(Sender: TObject);
begin
{while flag= false do  }

if flag=false then

Timer1.Enabled:=False;
PortController1.Close;
CloseFile(txt);
end;

procedure TForm1.ColorButton2Click(Sender: TObject);

begin
PortController1.Open('COM3','9600,8,N,1');
Timer1.Enabled:=tRUE;
ColorButton2.Enabled := False  ;
PortController1.ClearRQ;
rep:=0;
beep;
n:=0;
end;

procedure TForm1.Timer2Timer(Sender: TObject);
begin
  if a>47  then
begin

beep;
 end;
if a<15 then
 begin
beep;
end;
end;
end.
```

9 Delphi virtual component suppliers

There are many Delphi components available for use in virtual instrumentation projects, and they all rely on a simple variable byte value (0-255) which is passed to or from the component in the Delphi program. e.g. 'Component.Value ', a value of 255 will cause a meter movement to move to full deflection, a value of 127 will produce half scale deflection and 0 will keep the meter pointer at it's zero value. Conversely operating a control knob using 'componentknob.value', a movement of a quarter turn will output a value of 64, and a full turn 255, with the entire travel of the control knob outputting a value between 0 and 255. This chapter describes virtual components from several sources,but they all operate the same way .i.e. they all use a single byte to control operations. Some of the virtual components have properties which are infinitely variable and also have life-like movement such as a needle on a meter hitting the end stop and 'bouncing'. If you need to design a virtual instrument for a car ,aircraft or industrial plant all the necessary components can be found in this chapter.

9.1 GMS Active X components

GMS components are not installed as a Delphi package - which is the usual for a suite of instrument components, but rather as an active X control. Once a trial version is downloaded from the GMS website and installed on the PC, open the Delphi environment and click on 'Components', 'Import Active X Control' and install the required library. This includes angular gauges, automobile gauges, knobs, LED's, linear gauges and even aircraft light controls. To place a component on the Delphi Form, go to ActiveX in the top menu and the Icons for the component can then be pasted into the Form.

GMS Active X controls

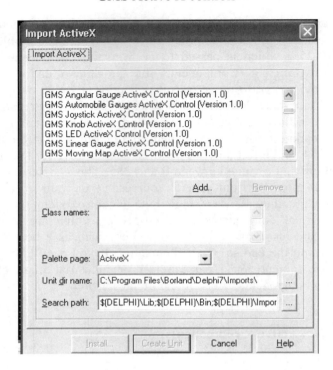

A working example of a car speedometer using a GMS virtual component is presented here. Microsoft's 44-pin Demo board is used as previously with the varying voltage on the potentiometer simulating a car toothed hall effect transducer and signal conditioning feeding the car speedometer. In this example a virtual LCD displays a digital readout of the speed and a car virtual analogue speedometer also displays the speed. The 0-5 volt voltage on Microchip's demo potentiometer is converted to a byte value by the onboard PIC and sent to the Delphi data variable for each component. i.e. Component.Value:

UniLCD1.Value:=data;
Car1.Value:=data;

Delphi Project CARX

This data value is then scaled in the program to produce a maximum speed of 250 mph on the speedometer scale. In addition to a speedometer, virtual components include a tachometer, fuel gauge, oil pressure gauge, water temperature gauge and a voltmeter:

The GMS car components

The Delphi Form for the car speedometer

The Delphi program for the virtual car speedometer can be found in Folder 'GMS',- file 'carx.pas'. It uses FTDI's D2XX unit for communications with the EVAL232R USB converter attached to the 44-pin demo board and so the DLL driver has to be loaded. All the properties of the speedometer component including scales, captions, fonts and captions can be customised by left-clicking on the speedometer component and/or editing its properties in the 'Object Inspector'.

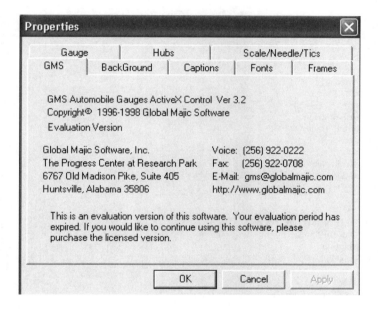

Delphi Program CAR.PAS

unit car;

interface

uses
  Windows, Messages, SysUtils, Classes, Graphics, Controls, Forms, Dialogs,
   MMSystem, StdCtrls, AbCBitBt, ExtCtrls,
  UniGraphicControl, UniVBSGraphicControl, UniVBSGraphicControlDA,
  UniMeter, UniAngularMeter, UniGauge, UniLCD, _GClass, AxCtrls, OleCtrls,
  DBOleCtl, CARLib_TLB;

type
  TForm1 = class(TForm)
    Timer1: TTimer;

```pascal
    AbColBitBtn1: TAbColBitBtn;
    UniLCD1: TUniLCD;
    Car1: TCar;

    procedure FormCreate(Sender: TObject);
    procedure Timer1Timer(Sender: TObject);
    procedure AbColBitBtn1Click(Sender: TObject);
  private
    { Private declarations }
  public
    { Public declarations }
  end;

var
  Form1: TForm1;
  txt:text;
   a:real;
   n:integer;
   data:byte;
   x:word;
   s:string;
   timestr:string;
  FT_Current_Baud:Dword;
   I:Integer;
   B :Byte;
      DevicePresent : Boolean;
  Selected_Device_Serial_Number : String;
  Selected_Device_Description : String;
const

    FT_DLL_NAME='ftd2xx.dll';
    FT_Out_Buffer_Size=1;
implementation
uses D2XXUnit;

{$R *.DFM}
procedure TForm1.FormCreate(Sender: TObject);
begin


UniLCD1.Value:=0;
Timer1.Interval:=1;
Timer1.Enabled:=True;
  Open_USB_Device;
  FT_Enable_Error_Report := false; // Disable Error Reporting
FT_Current_Baud:=FT_BAUD_9600;
FT_Current_DataBits:=FT_DATA_BITS_8;
FT_Current_StopBits:=FT_STOP_BITS_1;
FT_Current_Parity:=FT_PARITY_NONE;
I:= Set_USB_Device_BaudRate;
Set_USB_Device_RTS  ;
Set_USB_Device_DTR  ;

sndPlaySound('C:\system.wav',
  SND_NODEFAULT Or SND_ASYNC);
AssignFile(txt,'C:\DATAPOT\pot.txt');
```

```
    Rewrite(txt);
end;
procedure TForm1.Timer1Timer(Sender: TObject);
begin
Timer1.Enabled:=False;
Caption:=TimeToStr(Time)   ;
I:=Purge_USB_Device_In;
n:=0;

 I := Read_USB_Device_Buffer(1);
   data:=FT_In_Buffer[n];
asm
mov al,data
mov ah,0
mov x,ax
end;
a:=x;
 UniLCD1.Value:=data;
Car1.Value:=data;
beep;
timestr:=TimeToStr(Time);

 Str(a,s);
writeln(txt,a:3:1);

Str(a:2:1,s);
Caption:=TimeToStr(Time);
Canvas.TextOut(20,45,s);
Timer1.Enabled:=True;
end;
procedure TForm1.AbColBitBtn1Click(Sender: TObject);
begin
Timer1.Enabled:=False;
 Close_USB_Device;

CloseFile(txt);
close;
end;


end.
```

Also available from GMS are aircraft components:

And various gauges and meters and knobs:

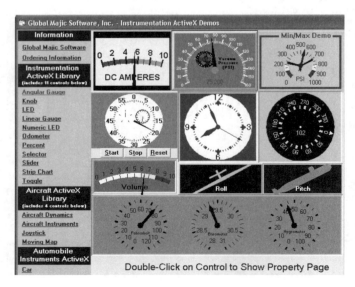

9.2 SDL Delphi Virtual components

SDL Project PRJMETER

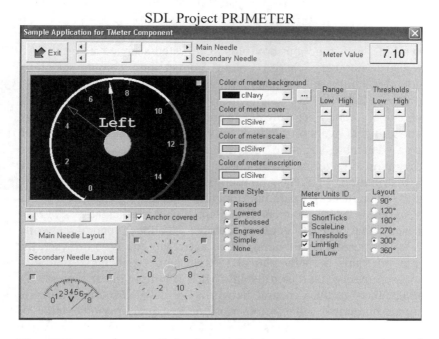

The SDL Component Suite is an industry leading collection of components supporting scientific and engineering computing. The entire suite consists of about 40 units covering a wide range of requirements in science and engineering. The component suite is available both for Delphi™ and for C++Builder™.

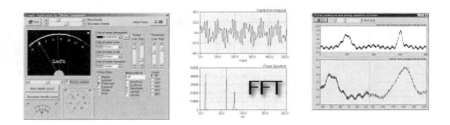

The following keywords cover only part of the SDL Component Suite and are intended to give you a quick impression of its power. :

3D rotation of data, 3D plot of data, associative arrays, atlas, atomic symbols, beta function, box plots, calibrated scanned images, CAS registry number, charts, colour conversion, colour selection, complex numbers, constants and conversion factors, contour plots, convex hull, CPU speed measurement, chemical structures, chemical data, chemical formulas, chi-square distribution, clustering, conversion routines, curve fitting, database of geographic data, datatable structure, determinant of a matrix, dendrograms, diagrams, directories, distributions, dot matrix labels, eigenvectors, F-distribution, FFT (fast Fourier transform), FIFO, box plots, file and disk access, gamma function, gauge, geographic atlas, geographic maps, gradient fill, grep, higher mathematics, HTML-based label, HTML entities, isotopes, Kohonen neural network, KNN (k-nearest neighbours), labels activating the Web browser, list view, maps, mathematical expression parsing, matrices, measurement data, meter display, MLR (multiple linear regression), molecular formulas, neural network, normal distribution, numeric input, numerical labels, numerical table editor, periodic table of elements, polar diagrams, PCA (principal component analysis), progress bars, quantiles of distributions, random generators, regression, RLE (run length encoding), rotation of data, scales, scientific chart, scrollable displays, slides, Smith chart, smoothing, splines, SOM, sorting of arrays, spread sheet, statistics, straight lines, streams, string array, string handling, surface plot of data, SVD (singular value decomposition), symbol selector, t-distribution, t-test, text tables, thumbnail images, universal save dialog, unique processor ID, valence electrons, wavelets, vectors, visualisation of data, VU meter

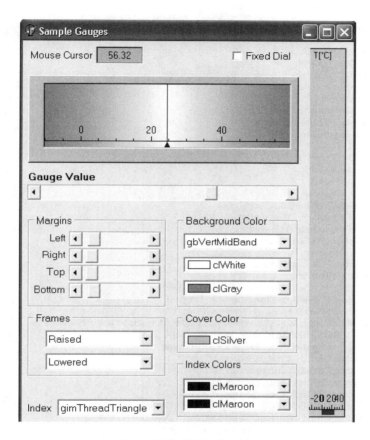

Example SDL Gauge

The software for the above example gauge can be found in Folder 'SDL', sub-folder 'Gaugesample'.

Program Gaugesample.
unit mainfrm;
interface
uses
　Windows, Messages, SysUtils, Variants, Classes, Graphics, Controls, Forms,
　Dialogs, SDL_gauge, StdCtrls, ExtCtrls, SDL_NumLab, SDL_Colsel, SDL_Scale;
type
　TForm1 = class(TForm)
　　SBarValue: TScrollBar;
　　NLabGaugeCoord: TNumLab;
　　CBoxIdxMode: TComboBox;
　　CBoxScaleFixed: TCheckBox;
　　SGauge1: TScaleGauge;
　　SGauge2: TScaleGauge;

```
    Label1: TLabel;
    GroupBox1: TGroupBox;
    ComboBox1: TComboBox;
    CSelBg1: TColSel;
    CSelBg2: TColSel;
    GroupBox2: TGroupBox;
    SBarMarginLeft: TScrollBar;
    SBarMarginRight: TScrollBar;
    SBarMarginTop: TScrollBar;
    SBarMarginBottom: TScrollBar;
    Label2: TLabel;
    Label3: TLabel;
    Label4: TLabel;
    Label5: TLabel;
    Label6: TLabel;
    GroupBox3: TGroupBox;
    CBoxFrStyleInner: TComboBox;
    CBoxFrStyleOuter: TComboBox;
    GroupBox4: TGroupBox;
    CSelColorCover: TColSel;
    GroupBox5: TGroupBox;
    CSelIdx1: TColSel;
    CSelIdx2: TColSel;
    procedure ComboBox1Change(Sender: TObject);
    procedure SBarValueChange(Sender: TObject);
    procedure Gauge1MouseMoveInGauge(Sender: TObject; InGauge: Boolean;
         Shift: TShiftState; rMousePos: Double);
    procedure CBoxIdxModeChange(Sender: TObject);
    procedure CSelIdx1Change(Sender: TObject);
    procedure CSelIdx2Change(Sender: TObject);
    procedure CSelBg1Change(Sender: TObject);
    procedure CSelBg2Change(Sender: TObject);
    procedure CBoxFrStyleOuterChange(Sender: TObject);
    procedure CBoxFrStyleInnerChange(Sender: TObject);
    procedure CBoxScaleFixedClick(Sender: TObject);
    procedure CSelColorCoverChange(Sender: TObject);
    procedure SBarMarginLeftChange(Sender: TObject);
    procedure SBarMarginRightChange(Sender: TObject);
    procedure SBarMarginTopChange(Sender: TObject);
    procedure SBarMarginBottomChange(Sender: TObject);
    procedure SGauge1MouseMoveInGauge(Sender: TObject; InGauge: Boolean;
      Shift: TShiftState; rMousePos: Double);
  private
  public
    { Public declarations }
  end;
var
  Form1: TForm1;
implementation
{$R *.dfm}
uses
  SDL_sdlbase;
procedure TForm1.ComboBox1Change(Sender: TObject);
begin
SGauge1.Background := TGaugeBackground(ComboBox1.ItemIndex);
end;
```

```
procedure TForm1.SBarValueChange(Sender: TObject);
begin
SGauge1.Value := SBarValue.Position/10;
SGauge2.Value := SBarValue.Position/10;
end;

procedure TForm1.Gauge1MouseMoveInGauge(Sender: TObject; InGauge: Boolean;
  Shift: TShiftState; rMousePos: Double);
begin
if InGauge
  then NLabGaugeCoord.ColorLabBakG := clGreen
  else NLabGaugeCoord.ColorLabBakG := clRed;
NLabGaugeCoord.Value := rMousePos;
end;
procedure TForm1.CBoxIdxModeChange(Sender: TObject);
begin
SGauge1.IndexMode := TGaugeIndexMode(CBoxIdxMode.ItemIndex);
end;
procedure TForm1.CSelIdx1Change(Sender: TObject);
begin
SGauge1.ColorIndex1 := CSelIdx1.SelColor;
end;
 procedure TForm1.CSelIdx2Change(Sender: TObject);
begin
SGauge1.ColorIndex2 := CSelIdx2.SelColor;
end;
procedure TForm1.CSelBg1Change(Sender: TObject);
begin
SGauge1.ColorBg1 := CSelBg1.SelColor;
end;
procedure TForm1.CSelBg2Change(Sender: TObject);
begin
SGauge1.ColorBg2 := CSelBG2.SelColor;
end;
procedure TForm1.CBoxFrStyleOuterChange(Sender: TObject);
begin
SGauge1.FrameStyleOuter := TFrameStyle(CBoxFrStyleOuter.ItemIndex);
end;

procedure TForm1.CBoxFrStyleInnerChange(Sender: TObject);
begin
SGauge1.FrameStyleInner := TFrameStyle(CBoxFrStyleInner.ItemIndex);
end;

procedure TForm1.CBoxScaleFixedClick(Sender: TObject);
begin
SGauge1.FixedDial := CBoxScaleFixed.Checked;
end;

procedure TForm1.CSelColorCoverChange(Sender: TObject);
begin
SGauge1.ColorCover := CSelColorCover.SelColor;
end;

procedure TForm1.SBarMarginLeftChange(Sender: TObject);
```

```
begin
SGauge1.MarginLeft := SBarMarginLeft.Position;
end;

procedure TForm1.SBarMarginRightChange(Sender: TObject);
begin
SGauge1.MarginRight := SBarMarginRight.Position;
end;

procedure TForm1.SBarMarginTopChange(Sender: TObject);
begin
SGauge1.MarginTop := SBarMarginTop.Position;
end;

procedure TForm1.SBarMarginBottomChange(Sender: TObject);
begin
SGauge1.MarginBottom := SBarMarginBottom.Position;
end;


(**************************************************************************
***********)
procedure TForm1.SGauge1MouseMoveInGauge(Sender: TObject; InGauge: Boolean;
  Shift: TShiftState; rMousePos: Double);
(**************************************************************************
***********)

begin
if InGauge
  then NLabGaugeCoord.ColorLabBakG := clLime
  else NLabGaugeCoord.ColorLabBakG := $6080FF;
NLabGaugeCoord.Value := rMousePos;
end;

end.
```

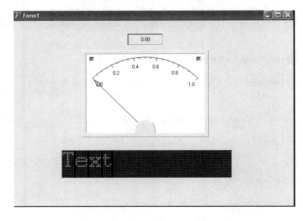

SDL Meter and Matrix display

9.3 Uniworks virtual components

Uniworks Technologies specialises in making high performance visual components for advanced graphic user interfaces, mainly for the scientific, automation and measurement fields. Realism is the keyword. UniComponents is a set of automation/measurement visual controls for major development platforms such as Microsoft Visual Studio, Visual Basic and Borland Delphi. It is also possible to integrate the controls in any COM-enabled environment such as Excel, Word or even webpages. There are a total of 5 component suites:
( ) Uniknob
( ) Unimatrix
( ) Unilcd
( ) Unicylinder Gauge
( ) Unimatrix

UNIKNOB
One may think OpenGL or GDI+ is involved. It is not, this is pure GDI at it's best. light source, shadows, texture and a whole set of other properties makes this knob component the most versatile available. A number of presets allow swift design, and detailed customisation is only clicks away. Properties for changing scale, font, marking, colour, texture and shape are available. The scale and ticks can be drawn onto the knob itself or on the surrounding panel. All graphics are rendered in real-time, no bitmaps are loaded nor stored within the component. The scale may be drawn either on the area surrounding the knob or on the knob itself, as can the ticks. The scale may also be alphanumeric in indexed mode as well as rotated either towards or from the knob itself. A property allows the scale values to be formatted to display any number of digits.

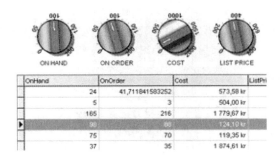

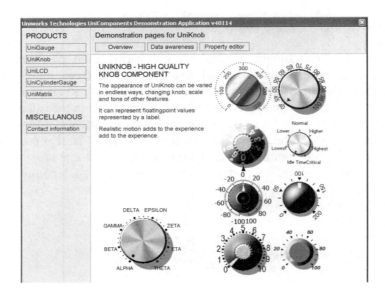

UNIMATRIX

UniMatrix is a visually superior alpha numeric matrix display component available in VCL and COM versions. Pixel width, height, spacing and several colouring properties allow customising the component to look like virtually any real-life matrix display. It also features a lightning effect that enhances the appearance. The text contents can be aligned to the left or right. Supported characters are numerical, uppercase and lowercase English characters and numerous symbols. The VCL version is also data aware, able to display the contents of any alphanumeric field as well as respond to content changes.

UNILCD

This is the best looking numeric LCD component available. UniLCD is Designed to look as realistic as possible, featuring display granularity, anti-aliasing, light-source and shadows as well as custom colouring of both display and symbols. Character features are managed using properties for LCD segment width and height. The display is rendered in high resolution and downsized for optimal graphic quality. The displayed value can be formatted to display an arbitrary number of decimals. The VCL version is also data aware and can be connected to any numerical data field to display its contents and even respond to content changes. The component can be aligned on a parent window. UniLCD is included in the UniComponents component library and is available in both VCL and COM versions, making it compatible with popular development platforms such as Microsoft Visual Studio, Borland Delphi and Borland C++Builder.

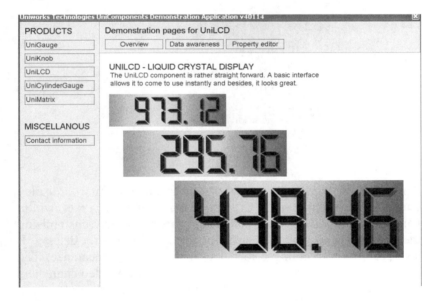

UNICYLINDER GAUGE

This component simulates a cylindrical gauge displaying data on a linear scale. It features true 3D, not plain shading like many similar products. Any font can be used for drawing the scale and the margin markings can be altered by setting number of markings, width, length and colour. The cylinder colour may also be changed. All graphics are real time rendered to prevent excessive memory usage. The VCL version of the component is data aware and may be connected to any numeric field in a database to display its contents and even respond to content changes.

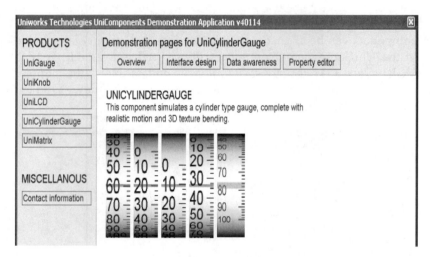

UNIMATRIX

UniMatrix is a visually superior alpha numeric matrix display component available in VCL and COM versions. Pixel width, height, spacing and several colouring properties allow customising the component to look like virtually any real-life matrix display. It also features a lightning effect that enhances the appearance. The text contents can be aligned to the left or right. Supported characters are numerical, uppercase and lowercase English characters and numerous symbols. The VCL version is also data aware, able to display the contents of any alphanumeric field as well as respond to content changes.

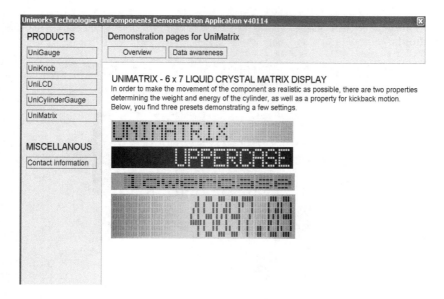

ANGULAR GAUGES

Instrumentation components no longer have to look like toys. This is, by far, the most advanced and visually complex gauge component available on the market. With more than 50 properties all forged into a powerful component editor, its appearance is highly customisable. All background graphics including ticks, scale and labels, are rendered in four times the size off screen and then downsized using anti-aliasing techniques for maximum graphic quality. Even the real time drawn needle may be anti aliased. The "tick" indicators have variable length, width, colour, start and end angles or could be excluded. This goes for the scale numbers that can be drawn using any font. The component can also display its current value using a numeric label.

Unlike most competing products, this component features real life needle movement. This means that the needle may slide smoothly into position instead of instantly and unnaturally change direction. If wanted, the application user can alter the gauge value by dragging the mouse over it, and if the component is connected to a datafield (VCL only) changes to the field are applied. This component can be applied in many automation instrumentation systems, including industry process control systems, aircraft instrumentation systems, remote monitor systems, scientific simulation systems and other fields.

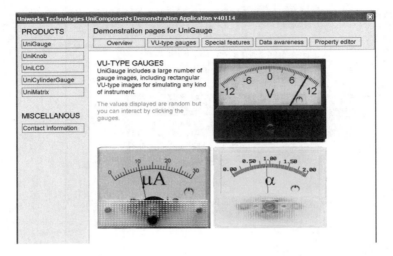

UNIGAUGE

A number of photo quality background image bitmaps are supplied with the component. These images are of a variety of gauge types, both 360° "circular" and 180° "VU" types, and they come in a range of styles and colours to suit different kinds of applications.

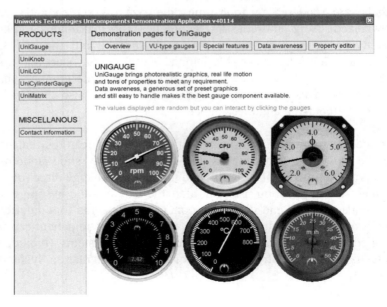

Program 'Project1' in Folder 'Uniworks Technologies' in subfolder 'Demos', provides an example how component 'UniGauge' is used.

```
Unit Unit1;

interface

uses
  Windows, Messages, SysUtils, Classes, Graphics, Controls, Forms, Dialogs,
  Db, Grids, DBGrids, DBTables, UniGraphicControl, UniVBSGraphicControl,
  UniVBSGraphicControlDA, UniMeter, UniAngularMeter, UniGauge;

type
  Tform1 = class(Tform)
    UniGauge1: TuniGauge;
    Table1: Ttable;
    DataSource1: TdataSource;
    DBGrid1: TDBGrid;
    Table1Length_In: TfloatField;
  private
    { Private declarations }
  public
    { Public declarations }
  end;

var
  Form1: Tform1;

implementation

{$R *.DFM}

end.
```

The Unigauge

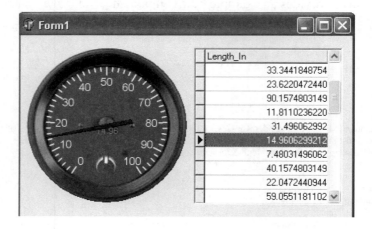

9.4 TMS Instrument workshop

TMS virtual components are infinitely adjustable and provide a comprehensive range in the following categories:

( )Meters
( )LED's
( )Switches
( )Scope panels

TMS Instrumentation Workshop is a comprehensive library full of components, methods and routines enabling you to create professional looking instrumentation and multimedia applications. A set containing over 80 instrumentation and digital components like leds, scopes, banners, sliders, knob controls, buttons, meters, panels with customized backgrounds and much more.:

 TVrAnalogClock: analogue clock in LCD style TVrAniButton: animated button using a bitmap filmstrip for animation TVrAnimate: animated image using a bitmap filmstrip for animation TVrArrow: arrow shaped button control TVrAngularMeter: rounded analogue meter device TVrBanner: scrolling bitmap with speed & direction control TVrBitmapButton: button using bitmap shape TVrBitmapCheckBox: checkbox with bitmaps for various states TVrBitmapDial: dial control using bitmap filmstrip for position display TVrBitmapImage: pattern bitmap image with various settings TVrBitmapList: container component for holding bitmaps TVrBitmapRadioButton: radiobutton with bitmaps for various states TVrBlinkLed: sizeable multi colour led with blink capability TVrBlotter: container control with child control placement management TVrBorder: bevelled outline TVrCalendar: control for selection of number or images from cells TVrCheckLed: checkbox with led TVrClock: timer display in LCD style TVrCompass: base class needle component TVrCopyFile: wrapper for file copy TVrCounter: counter display with customisable number bitmaps TVrDemoButton: 3D push button with various additional features TVrDeskTop: wallpaper TVrDigit: 7 segment LED ..etc

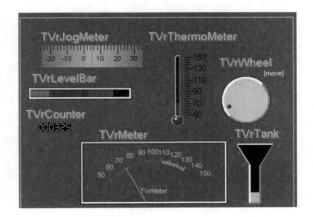

Meters

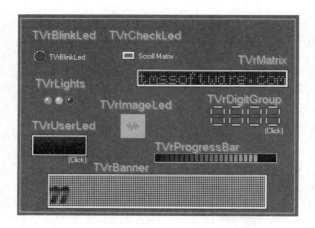

LED's

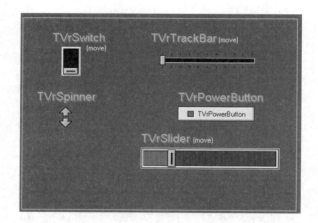

Switches

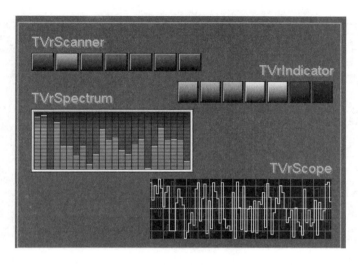

Scope Panel

Project 'ANGULAR' in Folder 'TIWDEMOS'

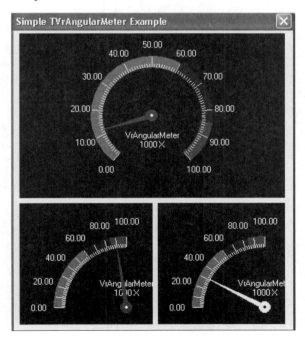

Delphi Program for ANGULAR.PAS

```
unit main;

interface

uses
  Windows, Messages, SysUtils, Classes, Graphics, Controls, Forms, Dialogs,
  VrControls, VrAngularMeter, VrThreads, VrLabel;

type
  TForm1 = class(TForm)
    VrAngularMeter1: TVrAngularMeter;
    VrAngularMeter2: TVrAngularMeter;
    VrAngularMeter3: TVrAngularMeter;
    VrTimer1: TVrTimer;
    procedure VrTimer1Timer(Sender: TObject);
  private
    { Private declarations }
  public
    { Public declarations }
  end;

var
  Form1: TForm1;

implementation

{$R *.DFM}

procedure TForm1.VrTimer1Timer(Sender: TObject);
begin
  VrAngularMeter1.Position := 10 + Random(5);
  VrAngularMeter2.Position := 80 + Random(20);
  VrAngularMeter3.Position := 20 + Random(20);
end;

end.
```

Project 'Anibutton' in Folder 'TWIDEMOS'

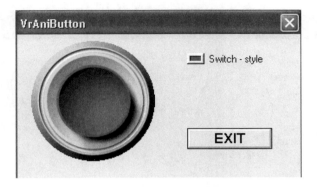

Delphi program for Anibutton.pas

```
unit uanib;
interface
uses
  Windows, Messages, SysUtils, Classes, Graphics, Controls, Forms, Dialogs,
  VrControls, VrBitmapDial, VrSystem, VrButtons, VrCheckLed;

type
 TForm1 = class(TForm)
   VrAniButton1: TVrAniButton;
   VrBitmapList1: TVrBitmapList;
   VrCheckLed1: TVrCheckLed;
   VrDemoButton1: TVrDemoButton;
   procedure VrCheckLed1Change(Sender: TObject);
   procedure VrDemoButton1Click(Sender: TObject);
  private
   { Private declarations }
  public
   { Public declarations }
  end;

var
 Form1: TForm1;
implementation
{$R *.DFM}
procedure TForm1.VrCheckLed1Change(Sender: TObject);
begin
  VrAniButton1.SwitchStyle := VrCheckLed1.Checked;
end;

procedure TForm1.VrDemoButton1Click(Sender: TObject);
begin
  Application.Terminate;
end;

end.
```

Project 'Rotary' in Folder 'TIWDEMO'

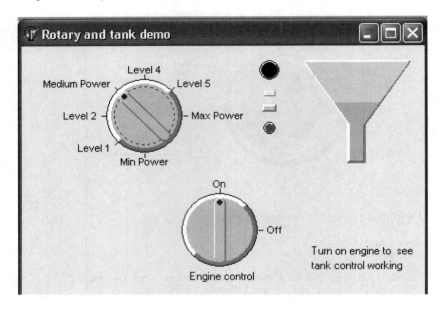

Program Rotary.PAS
unit urotary;

interface

uses
  Windows, Messages, SysUtils, Classes, Graphics, Controls, Forms, Dialogs,
  ComCtrls,mmsystem, VrControls, VrRotarySwitch, StdCtrls, VrBlinkLed,
  VrTank, ExtCtrls;

type
 TForm1 = class(TForm)
    VrRotarySwitch1: TVrRotarySwitch;
    VrTank1: TVrTank;
    VrBlinkLed1: TVrBlinkLed;
    VrBlinkLed2: TVrBlinkLed;
    VrBlinkLed3: TVrBlinkLed;
    VrBlinkLed4: TVrBlinkLed;
    Timer1: TTimer;
    VrRotarySwitch2: TVrRotarySwitch;
    Label1: TLabel;
    Label2: TLabel;
    Label3: TLabel;
    Label4: TLabel;
    procedure FormCreate(Sender: TObject);
    procedure VrRotarySwitch1Change(Sender: TObject);
    procedure Timer1Timer(Sender: TObject);

```
    procedure VrTank1OutofRange(Sender: TObject; var Min, Max,
      Pos: Double);
    procedure VrRotarySwitch2Change(Sender: TObject);
  private
    { Private declarations }
  public
    { Public declarations }
  end;

var
  Form1: TForm1;
  noRandom:integer;
implementation

{$R *.DFM}

procedure TForm1.FormCreate(Sender: TObject);
begin
  with VrRotarySwitch1.SwitchPositions do
  begin
    Add('Min Power');
    Add('Level 1');
    Add('Level 2');
    Add('Medium Power');
    Add('Level 4');
    Add('Level 5');
    Add('Max Power');
  end;
end;

procedure TForm1.VrRotarySwitch1Change(Sender: TObject);
begin
  mmsystem.sndPlaySound('sound.wav',SND_ASYNC);
  norandom := 100;
  case VrRotarySwitch1.SwitchPosition of
  0: begin
      VrBlinkLed1.EnableBlinking:=false;
      VrBlinkLed2.EnableBlinking:=false;
      VrBlinkLed3.EnableBlinking:=false;
      VrBlinkLed4.EnableBlinking:=false;
      norandom := 10;
    end;
  1:begin
      VrBlinkLed1.EnableBlinking:=true;
      VrBlinkLed2.EnableBlinking:=false;
      VrBlinkLed3.EnableBlinking:=false;
      VrBlinkLed4.EnableBlinking:=false;
      norandom := 20;
    end;
  2:begin
      VrBlinkLed1.EnableBlinking:=true;
      VrBlinkLed2.EnableBlinking:=true;
      VrBlinkLed3.EnableBlinking:=false;
      VrBlinkLed4.EnableBlinking:=false;
      norandom := 40;
    end;
```

```
   3:begin
       VrBlinkLed1.EnableBlinking:=true;
       VrBlinkLed2.EnableBlinking:=true;
       VrBlinkLed3.EnableBlinking:=true;
       VrBlinkLed4.EnableBlinking:=false;
       norandom := 60;
     end;
   4:begin
       VrBlinkLed1.EnableBlinking:=true;
       VrBlinkLed2.EnableBlinking:=true;
       VrBlinkLed3.EnableBlinking:=true;
       VrBlinkLed4.EnableBlinking:=true;
       norandom := 80;
     end;
   5: begin
       VrBlinkLed4.EnableBlinking:=false;
       norandom := 100;
     end;
   6: norandom:=1000;
   end;
end;
procedure TForm1.Timer1Timer(Sender: TObject);
begin
  if (VrTank1.Position < norandom) then
    VrTank1.Position := VrTank1.Position + random(2)
  else
  begin
    if (VrTank1.Position > 2) then
      VrTank1.Position := VrTank1.Position - random(2);
  end;
end;

procedure TForm1.VrTank1OutofRange(Sender: TObject; var Min, Max, Pos: Double);
begin
  label2.Visible := True;
  VrBlinkLed4.EnableBlinking := True;
end;

procedure TForm1.VrRotarySwitch2Change(Sender: TObject);
begin
  if VrRotarySwitch2.SwitchPosition = 1 then
  begin
    ShowMessage('Turned off');
    timer1.enabled := False;
  end
  else
  begin
    ShowMessage('Turned on');
    timer1.enabled := True;
  end;
end;

end.
```

Project 'Meterdemo' in Folder 'TIWDEMOS'

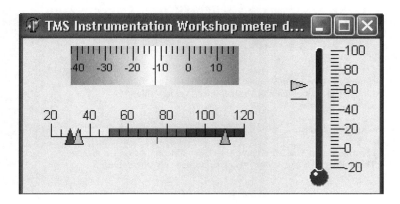

Program 'Meterdemo.pas'

```
unit UMeterDemo;

interface

uses
  Windows, Messages, SysUtils, Variants, Classes, Graphics, Controls, Forms,
  Dialogs, ExtCtrls, VrControls, VrJogMeter, VrThermoMeter, vrLineMeter;

type
  TForm1 = class(TForm)
    VrJogMeter1: TVrJogMeter;
    Timer1: TTimer;
    vrLineMeter1: TvrLineMeter;
    vrThermoMeter1: TvrThermoMeter;
    procedure Timer1Timer(Sender: TObject);
  private
    { Private declarations }
  public
    { Public declarations }
  end;

var
  Form1: TForm1;

implementation

{$R *.dfm}

procedure TForm1.Timer1Timer(Sender: TObject);
begin
  if random(20) > 10 then
    VrJogMeter1.Value.Value := VrJogMeter1.Value.Value + 1
  else
    VrJogMeter1.Value.Value := VrJogMeter1.Value.Value - 1;
```

```
if random(20) > 10 then
begin
  if VrLineMeter1.PeakMax.Value > VrLineMeter1.Value.Value then
    VrLineMeter1.Value.Value := VrLineMeter1.Value.Value + 1
end
else
begin
  if VrLineMeter1.PeakMin.Value < VrLineMeter1.Value.Value then
    VrLineMeter1.Value.Value := VrLineMeter1.Value.Value - 1;
end;

if random(20) > 10 then
begin
  if VrThermoMeter1.PeakMax.Value > VrThermoMeter1.Value.Value then
    VrThermoMeter1.Value.Value := VrThermoMeter1.Value.Value + 1
end
else
begin
  if VrThermoMeter1.PeakMin.Value < VrThermoMeter1.Value.Value then
    VrThermoMeter1.Value.Value := VrThermoMeter1.Value.Value - 1;
end
end;

end.
```

9.5 IOCOMP virtual instrument components

Iocomp's ActiveX/VCL Std Pack is a suite of 29 controls written for use in creating professional instrumentation applications using ActiveX or VCL development environments. These controls can be used for Scientific, Engineering, Medical, Oil and Gas, Semiconductor, Factory Automation, Aerospace, Military, Robotics, Telecommunications, Building and Home Automation, HMI, SCADA, and hundreds of other types of applications.

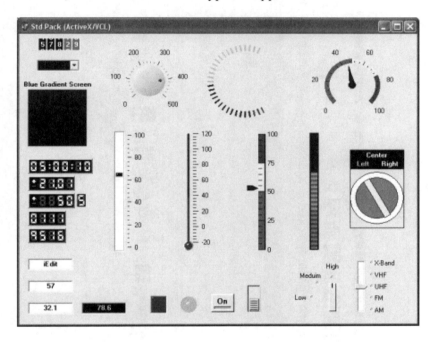

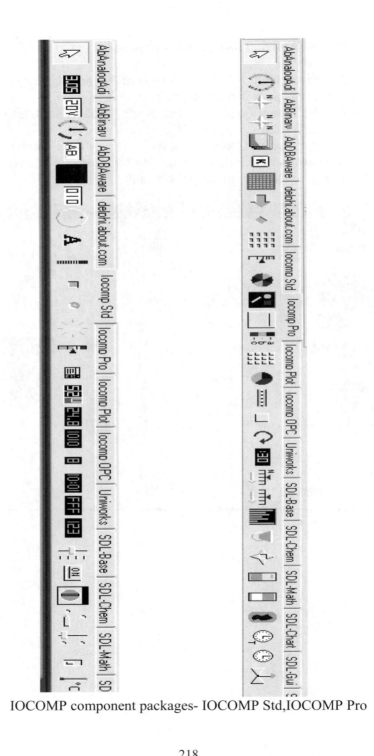

IOCOMP component packages- IOCOMP Std, IOCOMP Pro

All component properties are configurable by right-clicking on the component and invoking the Editor as shown below. Selecting 'Theme' in this example enables the customising of the gauge.

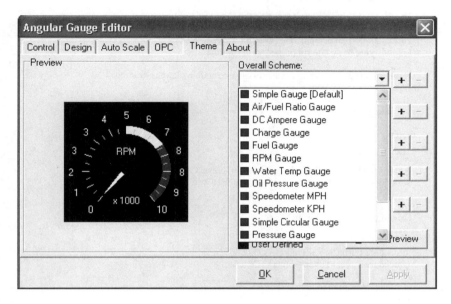

Project 'Compass' in Folder 'Demos'

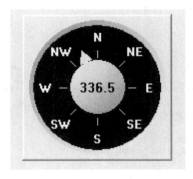

Program Compass.pass

```
//*********************************************
//
//Compass Control Example
```

```
//
//Copyright (c) 1997, 2008 Iocomp Software
//
//*******************************************************
//
//This simple example illustrates using the Compass Control.
//
//This example also demonstrates the following...
//1) Modifying basic control properties at runtime
//
//Permission is granted to distribute this source code without restriction.
//http://www.iocomp.com
//
unit CompassMain;

interface

uses
  Windows, Messages, SysUtils, Classes, Graphics, Controls, Forms, Dialogs,
  iComponent, iVCLComponent, iCompass, ExtCtrls, iTypes, iCustomComponent;

type
  TMainForm = class(TForm)
    Timer1: TTimer;
    iCompass1: TiCompass;
    procedure FormCreate(Sender: TObject);
    procedure Timer1Timer(Sender: TObject);
  private
    { Private declarations }
  public
    { Public declarations }
  end;

var
  MainForm: TMainForm;

implementation

{$R *.DFM}
//***************************************************************************
****
procedure TMainForm.FormCreate(Sender: TObject);
begin
  //Below shows a few of the many configuration
  //properties that can be set on the control.

  iCompass1.BorderStyle := ibsRaised;

  //Centering Control for demonstration purposes only.
  iCompass1.Left := MainForm.ClientWidth  div 2 - iCompass1.Width  div 2;
  iCompass1.Top  := MainForm.ClientHeight div 2 - iCompass1.Height div 2;
end;
//***************************************************************************
****
procedure TMainForm.Timer1Timer(Sender: TObject);
begin
```

```
    iCompass1.Direction := iCompass1.Direction + (Random(21) - 10)/2;

    //Used to keep compass setting between 0 - 360
    if (iCompass1.Direction <  0) then iCompass1.Direction := 360 + iCompass1.Direction;
    if (iCompass1.Direction > 360) then iCompass1.Direction := iCompass1.Direction - 360;
  end;
//********************************************************************
end.
```

9.6 CST software virtual components

Instrumentation Components family includes Studio for .NET, Studio for ActiveX, Studio for VCL, and AirGauge ActiveX. These components are high-quality and very popular in the process control, MMI, scientific, engineering, simulation, and data acquisition communities. All of them are designed for high speed.
You can download the evaluation edition from their web site and fully test the components in your project, they can be used freely for 60 days. The evaluation edition and the licensed edition are fully functional.

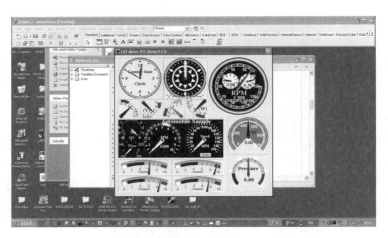

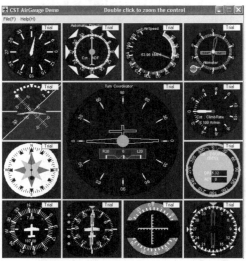

Project 'CST Odometer' in Folder 'Samples'

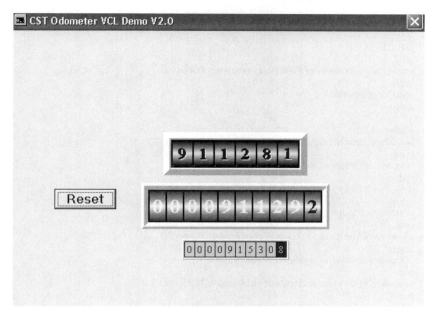

unit OdoDemo;

interface

uses
  Windows, Messages, SysUtils, Classes, Graphics, Controls, Forms, Dialogs,
  StdCtrls, ExtCtrls, jpeg, Odometer;

type
  TfrmOdometer = class(TForm)
    Timer1: TTimer;
    Timer3: TTimer;
    Timer2: TTimer;
    Odometer1: TOdometer;
    Odometer2: TOdometer;
    Odometer3: TOdometer;
    Button1: TButton;
    procedure FormDestroy(Sender: TObject);
    procedure Timer1Timer(Sender: TObject);
    procedure Timer2Timer(Sender: TObject);
    procedure Timer3Timer(Sender: TObject);
    procedure Button1Click(Sender: TObject);
  private

  public
    { Public declarations }
  end;

```
var
  frmOdometer: TfrmOdometer;

implementation

{$R *.DFM}

procedure TfrmOdometer.FormDestroy(Sender: TObject);
begin
  frmOdometer:=nil;
end;

//Sample Code;
procedure TfrmOdometer.Timer1Timer(Sender: TObject);
begin
  OdoMeter2.Value:=OdoMeter2.Value+0.1;
end;
procedure TfrmOdometer.Timer2Timer(Sender: TObject);
begin
  OdoMeter1.Value:=OdoMeter1.Value-0.1;
end;
procedure TfrmOdometer.Timer3Timer(Sender: TObject);
begin
  OdoMeter3.Value:=OdoMeter3.Value+1;
end;
procedure TfrmOdometer.Button1Click(Sender: TObject);
begin
  Odometer2.Reset;
end;

end.
```

CST Selector controls

10 Appendix

unit D2XXUnit;

interface
Uses Windows,Forms,Dialogs, Messages, SysUtils, Variants, Classes, Graphics, Controls;

Type FT_Result = Integer;

// Device Info Node structure for info list functions
type FT_Device_Info_Node = record
  Flags   : DWord;
  DeviceType   : Dword;
  ID   : DWord;
  LocID   : DWord;
  SerialNumber : array [0..15] of Char;
  Description   : array [0..63] of Char;
  DeviceHandle : DWord;
end;

type TDWordptr = ^DWord;

// Structure to hold EEPROM data for FT_EE_Program function
TFT_Program_Data = record
  Signature1   : DWord;
  Signature2   : DWord;
  Version   : DWord;
  VendorID   : Word;
  ProductID   : Word;
  Manufacturer   : PChar;
  ManufacturerID : PChar;
  Description   : PChar;
  SerialNumber   : PChar;
  MaxPower   : Word;
  PnP   : Word;
  SelfPowered   : Word;
  RemoteWakeup   : Word;
// Rev4 extensions
  Rev4   : Byte;
  IsoIn   : Byte;
  IsoOut   : Byte;
  PullDownEnable : Byte;
  SerNumEnable   : Byte;
  USBVersionEnable : Byte;
  USBVersion   : Word;
// FT2232C extensions
  Rev5   : Byte;
  IsoInA   : Byte;
  IsoInB   : Byte;
  IsoOutA   : Byte;
  IsoOutB   : Byte;
  PullDownEnable5 : Byte;

```
    SerNumEnable5      : Byte;
    USBVersionEnable5 : Byte;
    USBVersion5       : Word;
    AIsHighCurrent    : Byte;
    BIsHighCurrent    : Byte;
    IFAIsFifo         : Byte;
    IFAIsFifoTar      : Byte;
    IFAIsFastSer      : Byte;
    AIsVCP            : Byte;
    IFBIsFifo         : Byte;
    IFBIsFifoTar      : Byte;
    IFBIsFastSer      : Byte;
    BIsVCP            : Byte;
// FT232R extensions
    UseExtOsc         : Byte;
    HighDriveIOs      : Byte;
    EndpointSize      : Byte;
    PullDownEnableR   : Byte;
    SerNumEnableR     : Byte;
    InvertTXD         : Byte;
    InvertRXD         : Byte;
    InvertRTS         : Byte;
    InvertCTS         : Byte;
    InvertDTR         : Byte;
    InvertDSR         : Byte;
    InvertDCD         : Byte;
    InvertRI          : Byte;
    Cbus0             : Byte;
    Cbus1             : Byte;
    Cbus2             : Byte;
    Cbus3             : Byte;
    Cbus4             : Byte;
    RIsVCP            : Byte;
end;


// Exported Functions
// Classic Functions
Function GetFTDeviceCount : FT_Result;
Function GetFTDeviceSerialNo(DeviceIndex:DWord) : FT_Result;
Function GetFTDeviceDescription(DeviceIndex:DWord) : FT_Result;
Function GetFTDeviceLocation(DeviceIndex:DWord) : FT_Result;
Function Open_USB_Device : FT_Result;
Function Open_USB_Device_By_Serial_Number(Serial_Number:string) : FT_Result;
Function Open_USB_Device_By_Device_Description(Device_Description:string) : FT_Result;
Function Open_USB_Device_By_Device_Location(Location:DWord) : FT_Result;
Function Close_USB_Device : FT_Result;
Function Read_USB_Device_Buffer(Read_Count:Integer) : Integer;
Function Write_USB_Device_Buffer(Write_Count:Integer) : Integer;
Function Reset_USB_Device : FT_Result;
Function Set_USB_Device_BaudRate : FT_Result;
Function Set_USB_Device_BaudRate_Divisor(Divisor:Dword) : FT_Result;
Function Set_USB_Device_DataCharacteristics : FT_Result;
Function Set_USB_Device_FlowControl : FT_Result;
Function Set_USB_Device_RTS : FT_Result;
```

```
Function Clr_USB_Device_RTS : FT_Result;
Function Set_USB_Device_DTR : FT_Result;
Function Clr_USB_Device_DTR : FT_Result;
Function Get_USB_Device_ModemStatus : FT_Result;
Function Set_USB_Device_Chars : FT_Result;
Function Purge_USB_Device_Out : FT_Result;
Function Purge_USB_Device_In : FT_Result;
Function Set_USB_Device_TimeOuts(ReadTimeOut,WriteTimeOut:DWord) : FT_Result;
Function Get_USB_Device_QueueStatus : FT_Result;
Function Set_USB_Device_Break_On : FT_Result;
Function Set_USB_Device_Break_Off : FT_Result;
Function Get_USB_Device_Status : FT_Result;
Function Set_USB_Device_Event_Notification(EventMask:DWord) : FT_Result;
Function USB_FT_GetDeviceInfo(DevType,ID:DWord; SerialNumber,Description:array of char) : FT_Result;
Function Set_USB_Device_Reset_Pipe_Retry_Count(RetryCount:DWord) : FT_Result;
Function Stop_USB_Device_InTask : FT_Result;
Function Restart_USB_Device_InTask : FT_Result;
Function Reset_USB_Port : FT_Result;
Function Cycle_USB_Port : FT_Result;
Function Create_USB_Device_List : FT_Result;
Function Get_USB_Device_List : FT_Result;
Function Get_USB_Device_List_Detail(Index:DWord) : FT_Result;
// EEPROM Functions
function USB_FT_EE_Read : FT_Result;
function USB_FT_C_EE_Read : FT_Result;
function USB_FT_R_EE_Read : FT_Result;
function USB_FT_EE_Program : FT_Result;
function USB_FT_ReadEE(WordAddr:Dword) : FT_Result;
function USB_FT_WriteEE(WordAddr:Dword;WordData:Word) : FT_Result;
function USB_FT_EraseEE : FT_Result;
function USB_FT_EE_UARead : FT_Result;
function USB_FT_EE_UAWrite : FT_Result;
function USB_FT_EE_UASize : FT_Result;
// FT2232C, FT232BM and FT245BM Extended API Functions
Function Get_USB_Device_LatencyTimer : FT_Result;
Function Set_USB_Device_LatencyTimer(Latency : Byte) : FT_Result;
Function Get_USB_Device_BitMode(var BitMode:Byte) : FT_Result;
Function Set_USB_Device_BitMode(Mask, Enable:Byte) : FT_Result;
Function Set_USB_Parameters(InSize, OutSize:Dword) : FT_Result;

Function Get_USB_Driver_Version(DrVersion :  TDWordptr): FT_Result;
Function Get_USB_Library_Version(LbVersion :  TDWordptr): FT_Result;



Var
// Port Handle Returned by the Open Function
// Used by the Subsequent Function Calls
   FT_HANDLE : DWord = 0;
// Used to handle multiple device instances in future
// versions. Must be set to 0 for now.
//   PV_Device : DWord = 0;

// Holding Variables for the current settings
```

```
// Can be configured visually using the CFGUnit Unit
// or manually before calling SetUp_USB_Device
   FT_Current_Baud : Dword;
   FT_Current_DataBits : Byte;
   FT_Current_StopBits : Byte;
   FT_Current_Parity : Byte;
   FT_Current_FlowControl : Word;
   FT_RTS_On : Boolean;
   FT_DTR_On : Boolean;
   FT_Event_On : Boolean;
   FT_Error_On : Boolean;
   FT_XON_Value : Byte = $11;
   FT_XOFF_Value : Byte = $13;
   FT_EVENT_Value : Byte = $0;
   FT_ERROR_Value : Byte = $0;
// Used by CFGUnit to flag a bad value
   FT_SetupError : Boolean;
// Used to Return the current Modem Status
   FT_Modem_Status : DWord;
// Used to return the number of bytes pending
// in the Rx Buffer Queue
   FT_Q_Bytes : DWord;
   FT_TxQ_Bytes : DWord;
   FT_Event_Status : DWord;
// Used to Enable / Disable the Error Report Dialog
   FT_Enable_Error_Report : Boolean = True;
// Deposit for Get latency timer
   FT_LatencyRd : Byte;

   FT_DeviceInfoList : array of FT_Device_Info_Node;

   Manufacturer: array [0..63] of char;
   ManufacturerID: array [0..15] of char;
   Description: array [0..63] of char;
   SerialNumber: array [0..15] of char;
   LocID : DWord;
   EEDataBuffer : TFT_Program_Data;
   UserData : array [0..63] of byte;
   FT_UA_Size : integer;
   WordRead : Word;


Const
// FT_Result Values
   FT_OK = 0;
   FT_INVALID_HANDLE = 1;
   FT_DEVICE_NOT_FOUND = 2;
   FT_DEVICE_NOT_OPENED = 3;
   FT_IO_ERROR = 4;
   FT_INSUFFICIENT_RESOURCES = 5;
   FT_INVALID_PARAMETER = 6;
   FT_SUCCESS = FT_OK;
   FT_INVALID_BAUD_RATE = 7;
   FT_DEVICE_NOT_OPENED_FOR_ERASE = 8;
   FT_DEVICE_NOT_OPENED_FOR_WRITE = 9;
   FT_FAILED_TO_WRITE_DEVICE = 10;
```

```
    FT_EEPROM_READ_FAILED = 11;
    FT_EEPROM_WRITE_FAILED = 12;
    FT_EEPROM_ERASE_FAILED = 13;
    FT_EEPROM_NOT_PRESENT = 14;
    FT_EEPROM_NOT_PROGRAMMED = 15;
    FT_INVALID_ARGS = 16;
    FT_OTHER_ERROR = 17;
// FT_Open_Ex Flags
    FT_OPEN_BY_SERIAL_NUMBER = 1;
    FT_OPEN_BY_DESCRIPTION = 2;
    FT_OPEN_BY_LOCATION = 4;
// FT_List_Devices Flags
    FT_LIST_NUMBER_ONLY = $80000000;
    FT_LIST_BY_INDEX = $40000000;
    FT_LIST_ALL = $20000000;
// Baud Rate Selection
    FT_BAUD_300 = 300;
    FT_BAUD_600 = 600;
    FT_BAUD_1200 = 1200;
    FT_BAUD_2400 = 2400;
    FT_BAUD_4800 = 4800;
    FT_BAUD_9600 = 9600;
    FT_BAUD_14400 = 14400;
    FT_BAUD_19200 = 19200;
    FT_BAUD_38400 = 38400;
    FT_BAUD_57600 = 57600;
    FT_BAUD_115200 = 115200;
    FT_BAUD_230400 = 230400;
    FT_BAUD_460800 = 460800;
    FT_BAUD_921600 = 921600;
// Data Bits Selection
    FT_DATA_BITS_7 = 7;
    FT_DATA_BITS_8 = 8;
// Stop Bits Selection
    FT_STOP_BITS_1 = 0;
    FT_STOP_BITS_2 = 2;
// Parity Selection
    FT_PARITY_NONE = 0;
    FT_PARITY_ODD = 1;
    FT_PARITY_EVEN = 2;
    FT_PARITY_MARK = 3;
    FT_PARITY_SPACE = 4;
// Flow Control Selection
    FT_FLOW_NONE = $0000;
    FT_FLOW_RTS_CTS = $0100;
    FT_FLOW_DTR_DSR = $0200;
    FT_FLOW_XON_XOFF = $0400;
// Purge Commands
    FT_PURGE_RX = 1;
    FT_PURGE_TX = 2;
// Notification Events
    FT_EVENT_RXCHAR = 1;
    FT_EVENT_MODEM_STATUS = 2;
// Modem Status
    CTS = $10;
    DSR = $20;
```

```
    RI = $40;
    DCD = $80;


  // IO Buffer Sizes
    FT_In_Buffer_Size = $10000;    // 64k
    FT_In_Buffer_Index = FT_In_Buffer_Size - 1;
    FT_Out_Buffer_Size = $10000;   // 64k
    FT_Out_Buffer_Index = FT_Out_Buffer_Size - 1;
  // DLL Name
    FT_DLL_Name = 'FTD2XX.DLL';


  var
  // Declare Input and Output Buffers
    FT_In_Buffer : Array[0..FT_In_Buffer_Index] of Byte;
    FT_Out_Buffer : Array[0..FT_Out_Buffer_Index] of Byte;
  // A variable used to detect time-outs
  // Attach a timer to the main project form
  // which decrements this every 10mS if
  // FT_TimeOut_Count <> 0
    FT_TimeOut_Count : Integer = 0;
  // Used to determine how many bytes were
  // actually received by FT_Read_Device_All
  // in the case of a time-out
    FT_All_Bytes_Received : Integer = 0;
    FT_IO_Status : Ft_Result = FT_OK;
  // Used By FT_ListDevices
    FT_Device_Count : DWord;
    FT_Device_String_Buffer : array [1..50] of Char;
    FT_Device_String : String;
    FT_Device_Location : DWord;
    USB_Device_Info_Node : FT_Device_Info_Node;
    FT_Event_Handle : DWord;


  implementation
  //Classic functions
  function FT_GetNumDevices(pvArg1:Pointer; pvArg2:Pointer; dwFlags:Dword):FT_Result;
  stdcall; External FT_DLL_Name name 'FT_ListDevices';
  function FT_ListDevices(pvArg1:Dword; pvArg2:Pointer; dwFlags:Dword):FT_Result;
  stdcall; External FT_DLL_Name name 'FT_ListDevices';
  function FT_Open(Index:Integer; ftHandle:Pointer):FT_Result; stdcall; External
  FT_DLL_Name name 'FT_Open';
  function FT_OpenEx(pvArg1:Pointer; dwFlags:Dword; ftHandle:Pointer):FT_Result; stdcall;
  External FT_DLL_Name name 'FT_OpenEx';
  function FT_OpenByLocation(pvArg1:DWord; dwFlags:Dword;
  ftHandle:Pointer):FT_Result; stdcall; External FT_DLL_Name name 'FT_OpenEx';
  function FT_Close(ftHandle:Dword):FT_Result; stdcall; External FT_DLL_Name name
  'FT_Close';
  function FT_Read(ftHandle:Dword; FTInBuf:Pointer; BufferSize:LongInt;
  ResultPtr:Pointer):FT_Result; stdcall; External FT_DLL_Name name 'FT_Read';
  function FT_Write(ftHandle:Dword; FTOutBuf:Pointer; BufferSize:LongInt;
  ResultPtr:Pointer):FT_Result; stdcall; External FT_DLL_Name name 'FT_Write';
  function FT_ResetDevice(ftHandle:Dword):FT_Result; stdcall; External FT_DLL_Name
  name 'FT_ResetDevice';
```

function FT_SetBaudRate(ftHandle:Dword; BaudRate:DWord):FT_Result; stdcall; External FT_DLL_Name name 'FT_SetBaudRate';
function FT_SetDivisor(ftHandle:Dword; Divisor:DWord):FT_Result; stdcall; External FT_DLL_Name name 'FT_SetDivisor';
function FT_SetDataCharacteristics(ftHandle:Dword; WordLength,StopBits,Parity:Byte):FT_Result; stdcall; External FT_DLL_Name name 'FT_SetDataCharacteristics';
function FT_SetFlowControl(ftHandle:Dword; FlowControl:Word; XonChar,XoffChar:Byte):FT_Result; stdcall; External FT_DLL_Name name 'FT_SetFlowControl';
function FT_SetDtr(ftHandle:Dword):FT_Result; stdcall; External FT_DLL_Name name 'FT_SetDtr';
function FT_ClrDtr(ftHandle:Dword):FT_Result; stdcall; External FT_DLL_Name name 'FT_ClrDtr';
function FT_SetRts(ftHandle:Dword):FT_Result; stdcall; External FT_DLL_Name name 'FT_SetRts';
function FT_ClrRts(ftHandle:Dword):FT_Result; stdcall; External FT_DLL_Name name 'FT_ClrRts';
function FT_GetModemStatus(ftHandle:Dword; ModemStatus:Pointer):FT_Result; stdcall; External FT_DLL_Name name 'FT_GetModemStatus';
function FT_SetChars(ftHandle:Dword; EventChar,EventCharEnabled,ErrorChar,ErrorCharEnabled:Byte):FT_Result; stdcall; External FT_DLL_Name name 'FT_SetChars';
function FT_Purge(ftHandle:Dword; Mask:Dword):FT_Result; stdcall; External FT_DLL_Name name 'FT_Purge';
function FT_SetTimeouts(ftHandle:Dword; ReadTimeout,WriteTimeout:Dword):FT_Result; stdcall; External FT_DLL_Name name 'FT_SetTimeouts';
function FT_GetQueueStatus(ftHandle:Dword; RxBytes:Pointer):FT_Result; stdcall; External FT_DLL_Name name 'FT_GetQueueStatus';
function FT_SetBreakOn(ftHandle:Dword) : FT_Result; stdcall; External FT_DLL_Name name 'FT_SetBreakOn';
function FT_SetBreakOff(ftHandle:Dword) : FT_Result; stdcall; External FT_DLL_Name name 'FT_SetBreakOff';
function FT_GetStatus(ftHandle:DWord; RxBytes,TxBytes,EventStatus:Pointer):FT_Result; stdcall; External FT_DLL_Name name 'FT_GetStatus';
function FT_SetEventNotification(ftHandle:DWord; EventMask:DWord; pvArgs:Dword):FT_Result; stdcall; External FT_DLL_Name name 'FT_SetEventNotification';
function FT_GetDeviceInfo(ftHandle:DWord; DevType,ID,SerNum,Desc,pvDummy:Pointer) : FT_Result; stdcall; External FT_DLL_Name name 'FT_GetDeviceInfo';
function FT_SetResetPipeRetryCount(ftHandle:Dword; RetryCount:Dword):FT_Result; stdcall; External FT_DLL_Name name 'FT_SetResetPipeRetryCount';
function FT_StopInTask(ftHandle:Dword) : FT_Result; stdcall; External FT_DLL_Name name 'FT_StopInTask';
function FT_RestartInTask(ftHandle:Dword) : FT_Result; stdcall; External FT_DLL_Name name 'FT_RestartInTask';
function FT_ResetPort(ftHandle:Dword) : FT_Result; stdcall; External FT_DLL_Name name 'FT_ResetPort';
function FT_CyclePort(ftHandle:Dword) : FT_Result; stdcall; External 'FTD2XX.DLL' name 'FT_CyclePort';
function FT_CreateDeviceInfoList(NumDevs:Pointer):FT_Result; stdcall; External FT_DLL_Name name 'FT_CreateDeviceInfoList';
function FT_GetDeviceInfoList(pFT_Device_Info_List:Pointer; NumDevs:Pointer):FT_Result; stdcall; External FT_DLL_Name name 'FT_GetDeviceInfoList';

```
function FT_GetDeviceInfoDetail(Index:DWord;
Flags,DevType,ID,LocID,SerialNumber,Description,DevHandle:Pointer):FT_Result; stdcall;
External FT_DLL_Name name 'FT_GetDeviceInfoDetail';
function FT_GetDriverVersion(ftHandle:Dword; DrVersion:Pointer):FT_Result; stdcall;
External FT_DLL_Name name 'FT_GetDriverVersion';
function FT_GetLibraryVersion(LbVersion:Pointer):FT_Result; stdcall; External
FT_DLL_Name name 'FT_GetLibraryVersion';

// EEPROM functions
function FT_EE_Read(ftHandle:DWord; pEEData:Pointer):FT_Result; stdcall; External
FT_DLL_Name name 'FT_EE_Read';
function FT_EE_Program(ftHandle:DWord; pEEData:Pointer):FT_Result; stdcall; External
FT_DLL_Name name 'FT_EE_Program';
// EEPROM primitives - you need an NDA for EEPROM checksum
function FT_ReadEE(ftHandle:DWord; WordAddr:DWord; WordRead:Pointer):FT_Result;
stdcall; External FT_DLL_Name name 'FT_ReadEE';
function FT_WriteEE(ftHandle:DWord; WordAddr:DWord; WordData:word):FT_Result;
stdcall; External FT_DLL_Name name 'FT_WriteEE';
function FT_EraseEE(ftHandle:DWord):FT_Result; stdcall; External FT_DLL_Name name
'FT_EraseEE';
function FT_EE_UARead(ftHandle:DWord; Data:Pointer; DataLen:DWord;
BytesRead:Pointer):FT_Result; stdcall; External FT_DLL_Name name 'FT_EE_UARead';
function FT_EE_UAWrite(ftHandle:DWord; Data:Pointer; DataLen:DWord):FT_Result;
stdcall; External FT_DLL_Name name 'FT_EE_UAWrite';
function FT_EE_UASize(ftHandle:DWord; UASize:Pointer):FT_Result; stdcall; External
FT_DLL_Name name 'FT_EE_UASize';

// FT2232C, FT232BM and FT245BM Extended API Functions
function FT_GetLatencyTimer(ftHandle:Dword; Latency:Pointer):FT_Result; stdcall;
External FT_DLL_Name name 'FT_GetLatencyTimer';
function FT_SetLatencyTimer(ftHandle:Dword; Latency:Byte):FT_Result; stdcall; External
FT_DLL_Name name 'FT_SetLatencyTimer';
function FT_GetBitMode(ftHandle:Dword; BitMode:Pointer):FT_Result; stdcall; External
FT_DLL_Name name 'FT_GetBitMode';
function FT_SetBitMode(ftHandle:Dword; Mask,Enable:Byte):FT_Result; stdcall; External
FT_DLL_Name name 'FT_SetBitMode';
function FT_SetUSBParameters(ftHandle:Dword; InSize,OutSize:Dword):FT_Result; stdcall;
External FT_DLL_Name name 'FT_SetUSBParameters';


Procedure FT_Error_Report(ErrStr: String; PortStatus : Integer);
Var Str : String;
Begin
If Not FT_Enable_Error_Report then Exit;
If PortStatus = FT_OK then Exit;
Case PortStatus of
    FT_INVALID_HANDLE : Str := ErrStr+' - Invalid handle...';
    FT_DEVICE_NOT_FOUND : Str := ErrStr+' - Device not found...';
    FT_DEVICE_NOT_OPENED : Str := ErrStr+' - Device not opened...';
    FT_IO_ERROR : Str := ErrStr+' - General IO error...';
    FT_INSUFFICIENT_RESOURCES : Str := ErrStr+' - Insufficient resources...';
    FT_INVALID_PARAMETER : Str := ErrStr+' - Invalid parameter...';
    FT_INVALID_BAUD_RATE : Str := ErrStr+' - Invalid baud rate...';
    FT_DEVICE_NOT_OPENED_FOR_ERASE : Str := ErrStr+' Device not opened for
erase...';
```

```
    FT_DEVICE_NOT_OPENED_FOR_WRITE : Str := ErrStr+' Device not opened for
write...';
    FT_FAILED_TO_WRITE_DEVICE : Str := ErrStr+' - Failed to write...';
    FT_EEPROM_READ_FAILED : Str := ErrStr+' - EEPROM read failed...';
    FT_EEPROM_WRITE_FAILED : Str := ErrStr+' - EEPROM write failed...';
    FT_EEPROM_ERASE_FAILED : Str := ErrStr+' - EEPROM erase failed...';
    FT_EEPROM_NOT_PRESENT : Str := ErrStr+' - EEPROM not present...';
    FT_EEPROM_NOT_PROGRAMMED : Str := ErrStr+' - EEPROM not programmed...';
    FT_INVALID_ARGS : Str := ErrStr+' - Invalid arguments...';
    FT_OTHER_ERROR : Str := ErrStr+' - Other error ...';
  End;
  MessageDlg(Str, mtError, [mbOk], 0);
End;


Function GetDeviceString : String;
Var I : Integer;
Begin
 Result := ''; I := 1;
 FT_Device_String_Buffer[50] := Chr(0); // Just in case !
 While FT_Device_String_Buffer[I] <> Chr(0) do
  Begin
   Result := Result + FT_Device_String_Buffer[I];
   Inc(I);
  End;
End;


Procedure SetDeviceString ( S : String );
Var I,L : Integer;
Begin
 FT_Device_String_Buffer[1] := Chr(0);
 L := Length(S);  If L > 49 then L := 49;
 If L = 0 then Exit;
 For I := 1 to L do FT_Device_String_Buffer[I] := S[I];
 FT_Device_String_Buffer[L+1] := Chr(0);
End;


// FTD2XX functions from here
Function GetFTDeviceCount : FT_Result;
Begin
 Result := FT_GetNumDevices(@FT_Device_Count,Nil,FT_LIST_NUMBER_ONLY);
 If Result <> FT_OK then FT_Error_Report('GetFTDeviceCount',Result);
End;

Function GetFTDeviceSerialNo(DeviceIndex:DWord) : FT_Result;
Begin
 Result :=
FT_ListDevices(DeviceIndex,@SerialNumber,(FT_OPEN_BY_SERIAL_NUMBER or
FT_LIST_BY_INDEX));
 If Result = FT_OK then FT_Device_String := SerialNumber;
 If Result <> FT_OK then FT_Error_Report('GetFTDeviceSerialNo',Result);
End;
```

```
Function GetFTDeviceDescription(DeviceIndex:DWord) : FT_Result;
Begin
Result := FT_ListDevices(DeviceIndex,@Description,(FT_OPEN_BY_DESCRIPTION or
FT_LIST_BY_INDEX));
If Result = FT_OK then FT_Device_String := Description;
If Result <> FT_OK then FT_Error_Report('GetFTDeviceDescription',Result);
End;


Function GetFTDeviceLocation(DeviceIndex:DWord) : FT_Result;
Begin
Result := FT_ListDevices(DeviceIndex,@LocID,(FT_OPEN_BY_LOCATION or
FT_LIST_BY_INDEX));
If Result = FT_OK then FT_Device_Location := LocID;
If Result <> FT_OK then FT_Error_Report('GetFTDeviceLocation',Result);
End;


Function Open_USB_Device : FT_Result;
Var
 DevIndex : DWord;
Begin
DevIndex := 0;
Result := FT_Open(DevIndex,@FT_Handle);
If Result <> FT_OK then FT_Error_Report('FT_Open',Result);
End;


Function Open_USB_Device_By_Serial_Number(Serial_Number:string) : FT_Result;
Begin
SetDeviceString(Serial_Number);
Result :=
FT_OpenEx(@FT_Device_String_Buffer,FT_OPEN_BY_SERIAL_NUMBER,@FT_Handle
);
If Result <> FT_OK then FT_Error_Report('Open_USB_Device_By_Serial_Number',Result);
End;


Function Open_USB_Device_By_Device_Description(Device_Description:string) :
FT_Result;
Begin
SetDeviceString(Device_Description);
Result :=
FT_OpenEx(@FT_Device_String_Buffer,FT_OPEN_BY_DESCRIPTION,@FT_Handle);
If Result <> FT_OK then
FT_Error_Report('Open_USB_Device_By_Device_Description',Result);
End;


Function Open_USB_Device_By_Device_Location(Location:DWord) : FT_Result;
Begin
Result := FT_OpenByLocation(Location,FT_OPEN_BY_LOCATION,@FT_Handle);
If Result <> FT_OK then
FT_Error_Report('Open_USB_Device_By_Device_Location',Result);
End;
```

```pascal
Function Close_USB_Device : FT_Result;
Begin
Result := FT_Close(FT_Handle);
If Result <> FT_OK then FT_Error_Report('FT_Close',Result);
End;


function Read_USB_Device_Buffer( Read_Count : Integer ) : Integer;
// Reads Read_Count Bytes (or less) from the USB device to the FT_In_Buffer
// Function returns the number of bytes actually received  which may range from zero
// to the actual number of bytes requested, depending on how many have been received
// at the time of the request + the read timeout value.
Var Read_Result : Integer;
Begin

if (read_count = 1) then
  begin
  read_result := read_count;
  end;
FT_IO_Status := FT_Read(FT_Handle,@FT_In_Buffer,Read_Count,@Read_Result);
If FT_IO_Status <> FT_OK then FT_Error_Report('FT_Read',FT_IO_Status);
Result := Read_Result;
End;


function Write_USB_Device_Buffer( Write_Count : Integer ) : Integer;
// Writes Write_Count Bytes from FT_Out_Buffer to the USB device
// Function returns the number of bytes actually sent
// In this example, Write_Count should be 32k bytes max
Var Write_Result : Integer;
Begin
FT_IO_Status := FT_Write(FT_Handle,@FT_Out_Buffer,Write_Count,@Write_Result);
If FT_IO_Status <> FT_OK then FT_Error_Report('FT_Write',FT_IO_Status);
Result := Write_Result;
End;


Function Reset_USB_Device : FT_Result;
Begin
Result := FT_ResetDevice(FT_Handle);
If Result <> FT_OK then FT_Error_Report('FT_ResetDevice',Result);
End;


Function Set_USB_Device_BaudRate : FT_Result;
Begin
Result := FT_SetBaudRate(FT_Handle,FT_Current_Baud);
If Result <> FT_OK then FT_Error_Report('FT_SetBaudRate',Result);
End;


Function Set_USB_Device_BaudRate_Divisor(Divisor:Dword) : FT_Result;
Begin
Result := FT_SetDivisor(FT_Handle,Divisor);
If Result <> FT_OK then FT_Error_Report('FT_SetDivisor',Result);
```

End;

Function Set_USB_Device_DataCharacteristics : FT_Result;
Begin
Result :=
FT_SetDataCharacteristics(FT_Handle,FT_Current_DataBits,FT_Current_StopBits,FT_Current_Parity);
If Result <> FT_OK then FT_Error_Report('FT_SetDataCharacteristics',Result);
End;

Function Set_USB_Device_FlowControl : FT_Result;
Begin
Result :=
FT_SetFlowControl(FT_Handle,FT_Current_FlowControl,FT_XON_Value,FT_XOFF_Value);
If Result <> FT_OK then FT_Error_Report('FT_SetFlowControl',Result);
End;

Function Set_USB_Device_RTS : FT_Result;
Begin
Result := FT_SetRTS(FT_Handle);
If Result <> FT_OK then FT_Error_Report('FT_SetRTS',Result);
End;

Function Clr_USB_Device_RTS : FT_Result;
Begin
Result := FT_ClrRTS(FT_Handle);
If Result <> FT_OK then FT_Error_Report('FT_ClrRTS',Result);
End;

Function Set_USB_Device_DTR : FT_Result;
Begin
Result := FT_SetDTR(FT_Handle);
If Result <> FT_OK then FT_Error_Report('FT_SetDTR',Result);
End;

Function Clr_USB_Device_DTR : FT_Result;
Begin
Result := FT_ClrDTR(FT_Handle);
If Result <> FT_OK then FT_Error_Report('FT_ClrDTR',Result);
End;

Function Get_USB_Device_ModemStatus : FT_Result;
Begin
Result := FT_GetModemStatus(FT_Handle,@FT_Modem_Status);
If Result <> FT_OK then FT_Error_Report('FT_GetModemStatus',Result);
End;

```
Function Set_USB_Device_Chars : FT_Result;
Var Events_On,Errors_On : Byte;
Begin
If FT_Event_On then Events_On := 1 else Events_On := 0;
If FT_Error_On then Errors_On := 1 else Errors_On := 0;
Result :=
FT_SetChars(FT_Handle,FT_EVENT_Value,Events_On,FT_ERROR_Value,Errors_On);
If Result <> FT_OK then FT_Error_Report('FT_SetChars',Result);
End;


Function Purge_USB_Device_Out : FT_Result;
Begin
Result := FT_Purge(FT_Handle,FT_PURGE_RX);
If Result <> FT_OK then FT_Error_Report('FT_Purge RX',Result);
End;

Function Purge_USB_Device_In : FT_Result;
Begin
Result := FT_Purge(FT_Handle,FT_PURGE_TX);
If Result <> FT_OK then FT_Error_Report('FT_Purge TX',Result);
End;


Function Set_USB_Device_TimeOuts(ReadTimeOut,WriteTimeOut:DWord) : FT_Result;
Begin
Result := FT_SetTimeouts(FT_Handle,ReadTimeout,WriteTimeout);
If Result <> FT_OK then FT_Error_Report('FT_SetTimeouts',Result);
End;


Function Get_USB_Device_QueueStatus : FT_Result;
Begin
Result := FT_GetQueueStatus(FT_Handle,@FT_Q_Bytes);
If Result <> FT_OK then FT_Error_Report('FT_GetQueueStatus',Result);
End;


Function Set_USB_Device_Break_On : FT_Result;
Begin
Result := FT_SetBreakOn(FT_Handle);
If Result <> FT_OK then FT_Error_Report('FT_SetBreakOn',Result);
End;


Function Set_USB_Device_Break_Off : FT_Result;
Begin
Result := FT_SetBreakOff(FT_Handle);
If Result <> FT_OK then FT_Error_Report('FT_SetBreakOff',Result);
End;


Function Get_USB_Device_Status : FT_Result;
Begin
Result := FT_GetStatus(FT_Handle, @FT_Q_Bytes, @FT_TxQ_Bytes,
@FT_Event_Status);
```

```
If Result <> FT_OK then FT_Error_Report('FT_GetStatus',Result);
End;


Function Set_USB_Device_Event_Notification(EventMask:DWord) : FT_Result;
Begin
Result := FT_SetEventNotification(FT_Handle,EventMask,FT_Event_Handle);
If Result <> FT_OK then FT_Error_Report('FT_SetEventNotification ',Result);
End;


Function USB_FT_GetDeviceInfo(DevType,ID:DWord; SerialNumber,Description:array of char) : FT_Result;
begin
Result :=
FT_GetDeviceInfo(FT_Handle,@DevType,@ID,@SerialNumber,@Description,Nil);
If Result <> FT_OK then FT_Error_Report('FT_GetDeviceInfo ',Result);
end;


Function Set_USB_Device_Reset_Pipe_Retry_Count(RetryCount:DWord) : FT_Result;
Begin
Result := FT_SetResetPiperetryCount(FT_Handle, RetryCount);
If Result <> FT_OK then FT_Error_Report('FT_SetResetPipeRetryCount',Result);
End;


Function Stop_USB_Device_InTask : FT_Result;
Begin
Result := FT_StopInTask(FT_Handle);
If Result <> FT_OK then FT_Error_Report('FT_StopInTask',Result);
End;


Function Restart_USB_Device_InTask : FT_Result;
Begin
Result := FT_RestartInTask(FT_Handle);
If Result <> FT_OK then FT_Error_Report('FT_RestartInTask',Result);
End;


Function Reset_USB_Port : FT_Result;
Begin
Result := FT_ResetPort(FT_Handle);
If Result <> FT_OK then FT_Error_Report('FT_ResetPort',Result);
End;


Function Cycle_USB_Port : FT_Result;
Begin
Result := FT_CyclePort(FT_Handle);
If Result <> FT_OK then FT_Error_Report('FT_CyclePort',Result);
End;


Function Create_USB_Device_List : FT_Result;
```

```
Begin
Result := FT_CreateDeviceInfoList(@FT_Device_Count);
If Result <> FT_OK then FT_Error_Report('FT_CreateDeviceInfoList',Result);
End;


Function Get_USB_Device_List : FT_Result;
Begin
SetLength(FT_DeviceInfoList,FT_Device_Count);
Result := FT_GetDeviceInfoList(FT_DeviceInfoList, @FT_Device_Count);
If Result <> FT_OK then FT_Error_Report('FT_GetDeviceInfoList',Result);
End;

Function Get_USB_Driver_Version(DrVersion : TDWordPtr) : FT_Result;
Begin
  Result := FT_GetDriverVersion(FT_Handle, DrVersion);
  If Result <> FT_OK then FT_Error_Report('FT_GetDriverVersion',Result);
End;

Function Get_USB_Library_Version(LbVersion : TDWordPtr) : FT_Result;
Begin
  Result := FT_GetLibraryVersion(LbVersion);
  If Result <> FT_OK then FT_Error_Report('FT_GetLibraryVersion',Result);
End;


Function Get_USB_Device_List_Detail(Index:DWord) : FT_Result;
Begin
// Initialise structure
USB_Device_Info_Node.Flags := 0;
USB_Device_Info_Node.DeviceType := 0;
USB_Device_Info_Node.ID := 0;
USB_Device_Info_Node.LocID := 0;
USB_Device_Info_Node.SerialNumber := '';
USB_Device_Info_Node.Description := '';
USB_Device_Info_Node.DeviceHandle := 0;
Result :=
FT_GetDeviceInfoDetail(Index,@USB_Device_Info_Node.Flags,@USB_Device_Info_Node
.DeviceType,

@USB_Device_Info_Node.ID,@USB_Device_Info_Node.LocID,@USB_Device_Info_Node
.SerialNumber,
   (@USB_Device_Info_Node.Description,@USB_Device_Info_Node.DeviceHandle);
If Result <> FT_OK then FT_Error_Report('FT_GetDeviceInfoListDetail',Result);
End;



function USB_FT_EE_Read : FT_Result;
// Read BM/AM device EEPROM
begin
EEDataBuffer.Signature1 := 0;
EEDataBuffer.Signature2 := $FFFFFFFF;
EEDataBuffer.Version := 0;  // 0 for AM/BM, 1 for C, 2 for R
EEDataBuffer.VendorId :=0;
EEDataBuffer.ProductId := 0;
EEDataBuffer.Manufacturer := @Manufacturer;
```

```
EEDataBuffer.ManufacturerId := @ManufacturerId;
EEDataBuffer.Description := @Description;
EEDataBuffer.SerialNumber := @SerialNumber;
EEDataBuffer.MaxPower := 0;
EEDataBuffer.PnP := 0;
EEDataBuffer.SelfPowered := 0;
EEDataBuffer.RemoteWakeup := 0;
EEDataBuffer.Rev4 := 0;
EEDataBuffer.IsoIn := 0;
EEDataBuffer.IsoOut := 0;
EEDataBuffer.PullDownEnable := 0;
EEDataBuffer.SerNumEnable := 0;
EEDataBuffer.USBVersionEnable := 0;
EEDataBuffer.USBVersion := 0;
// FT2232C Extensions
EEDataBuffer.Rev5 := 0;
EEDataBuffer.IsoInA := 0;
EEDataBuffer.IsoInB := 0;
EEDataBuffer.IsoOutA := 0;
EEDataBuffer.IsoOutB := 0;
EEDataBuffer.PullDownEnable5 := 0;
EEDataBuffer.SerNumEnable5 := 0;
EEDataBuffer.USBVersionEnable5 := 0;
EEDataBuffer.USBVersion5 := 0;
EEDataBuffer.AIsHighCurrent := 0;
EEDataBuffer.BIsHighCurrent := 0;
EEDataBuffer.IFAIsFifo := 0;
EEDataBuffer.IFAIsFifoTar := 0;
EEDataBuffer.IFAIsFastSer := 0;
EEDataBuffer.AIsVCP := 0;
EEDataBuffer.IFBIsFifo := 0;
EEDataBuffer.IFBIsFifoTar := 0;
EEDataBuffer.IFBIsFastSer := 0;
EEDataBuffer.BIsVCP := 0;
// FT232R extensions
EEDataBuffer.UseExtOsc := 0;
EEDataBuffer.HighDriveIOs := 0;
EEDataBuffer.EndpointSize := 0;
EEDataBuffer.PullDownEnableR := 0;
EEDataBuffer.SerNumEnableR := 0;
EEDataBuffer.InvertTXD := 0;
EEDataBuffer.InvertRXD := 0;
EEDataBuffer.InvertRTS := 0;
EEDataBuffer.InvertCTS := 0;
EEDataBuffer.InvertDTR := 0;
EEDataBuffer.InvertDSR := 0;
EEDataBuffer.InvertDCD := 0;
EEDataBuffer.InvertRI := 0;
EEDataBuffer.Cbus0 := 0;
EEDataBuffer.Cbus1 := 0;
EEDataBuffer.Cbus2 := 0;
EEDataBuffer.Cbus3 := 0;
EEDataBuffer.Cbus4 := 0;
EEDataBuffer.RIsVCP := 0;
Result :=  FT_EE_Read(FT_Handle,@EEDataBuffer);
If Result <> FT_OK then FT_Error_Report('FT_EE_Read ',Result);
```

end;

function USB_FT_C_EE_Read : FT_Result;
// Read FT2232C device EEPROM
begin
EEDataBuffer.Signature1 := 0;
EEDataBuffer.Signature2 := $FFFFFFFF;
EEDataBuffer.Version := 1; // 0 for AM/BM, 1 for C, 2 for R
EEDataBuffer.VendorId :=0;
EEDataBuffer.ProductId := 0;
EEDataBuffer.Manufacturer := @Manufacturer;
EEDataBuffer.ManufacturerId := @ManufacturerId;
EEDataBuffer.Description := @Description;
EEDataBuffer.SerialNumber := @SerialNumber;
EEDataBuffer.MaxPower := 0;
EEDataBuffer.PnP := 0;
EEDataBuffer.SelfPowered := 0;
EEDataBuffer.RemoteWakeup := 0;
EEDataBuffer.Rev4 := 0;
EEDataBuffer.IsoIn := 0;
EEDataBuffer.IsoOut := 0;
EEDataBuffer.PullDownEnable := 0;
EEDataBuffer.SerNumEnable := 0;
EEDataBuffer.USBVersionEnable := 0;
EEDataBuffer.USBVersion := 0;
// FT2232C Extensions
EEDataBuffer.Rev5 := 0;
EEDataBuffer.IsoInA := 0;
EEDataBuffer.IsoInB := 0;
EEDataBuffer.IsoOutA := 0;
EEDataBuffer.IsoOutB := 0;
EEDataBuffer.PullDownEnable5 := 0;
EEDataBuffer.SerNumEnable5 := 0;
EEDataBuffer.USBVersionEnable5 := 0;
EEDataBuffer.USBVersion5 := 0;
EEDataBuffer.AIsHighCurrent := 0;
EEDataBuffer.BIsHighCurrent := 0;
EEDataBuffer.IFAIsFifo := 0;
EEDataBuffer.IFAIsFifoTar := 0;
EEDataBuffer.IFAIsFastSer := 0;
EEDataBuffer.AIsVCP := 0;
EEDataBuffer.IFBIsFifo := 0;
EEDataBuffer.IFBIsFifoTar := 0;
EEDataBuffer.IFBIsFastSer := 0;
EEDataBuffer.BIsVCP := 0;
// FT232R extensions
EEDataBuffer.UseExtOsc := 0;
EEDataBuffer.HighDriveIOs := 0;
EEDataBuffer.EndpointSize := 0;
EEDataBuffer.PullDownEnableR := 0;
EEDataBuffer.SerNumEnableR := 0;
EEDataBuffer.InvertTXD := 0;
EEDataBuffer.InvertRXD := 0;
EEDataBuffer.InvertRTS := 0;
EEDataBuffer.InvertCTS := 0;

```
EEDataBuffer.InvertDTR := 0;
EEDataBuffer.InvertDSR := 0;
EEDataBuffer.InvertDCD := 0;
EEDataBuffer.InvertRI := 0;
EEDataBuffer.Cbus0 := 0;
EEDataBuffer.Cbus1 := 0;
EEDataBuffer.Cbus2 := 0;
EEDataBuffer.Cbus3 := 0;
EEDataBuffer.Cbus4 := 0;
EEDataBuffer.RIsVCP := 0;
Result :=  FT_EE_Read(FT_Handle,@EEDataBuffer);
If Result <> FT_OK then FT_Error_Report('FT_EE_Read ',Result);
end;


function USB_FT_R_EE_Read : FT_Result;
// Read FT232R device EEPROM
begin
EEDataBuffer.Signature1 := 0;
EEDataBuffer.Signature2 := $FFFFFFFF;
EEDataBuffer.Version := 2;  // 0 for AM/BM, 1 for C, 2 for R
EEDataBuffer.VendorId :=0;
EEDataBuffer.ProductId := 0;
EEDataBuffer.Manufacturer := @Manufacturer;
EEDataBuffer.ManufacturerId := @ManufacturerId;
EEDataBuffer.Description := @Description;
EEDataBuffer.SerialNumber := @SerialNumber;
EEDataBuffer.MaxPower := 0;
EEDataBuffer.PnP := 0;
EEDataBuffer.SelfPowered := 0;
EEDataBuffer.RemoteWakeup := 0;
EEDataBuffer.Rev4 := 0;
EEDataBuffer.IsoIn := 0;
EEDataBuffer.IsoOut := 0;
EEDataBuffer.PullDownEnable := 0;
EEDataBuffer.SerNumEnable := 0;
EEDataBuffer.USBVersionEnable := 0;
EEDataBuffer.USBVersion := 0;
// FT2232C Extensions
EEDataBuffer.Rev5 := 0;
EEDataBuffer.IsoInA := 0;
EEDataBuffer.IsoInB := 0;
EEDataBuffer.IsoOutA := 0;
EEDataBuffer.IsoOutB := 0;
EEDataBuffer.PullDownEnable5 := 0;
EEDataBuffer.SerNumEnable5 := 0;
EEDataBuffer.USBVersionEnable5 := 0;
EEDataBuffer.USBVersion5 := 0;
EEDataBuffer.AIsHighCurrent := 0;
EEDataBuffer.BIsHighCurrent := 0;
EEDataBuffer.IFAIsFifo := 0;
EEDataBuffer.IFAIsFifoTar := 0;
EEDataBuffer.IFAIsFastSer := 0;
EEDataBuffer.AIsVCP := 0;
EEDataBuffer.IFBIsFifo := 0;
EEDataBuffer.IFBIsFifoTar := 0;
```

```
EEDataBuffer.IFBIsFastSer := 0;
EEDataBuffer.BIsVCP := 0;
// FT232R extensions
EEDataBuffer.UseExtOsc := 0;
EEDataBuffer.HighDriveIOs := 0;
EEDataBuffer.EndpointSize := 0;
EEDataBuffer.PullDownEnableR := 0;
EEDataBuffer.SerNumEnableR := 0;
EEDataBuffer.InvertTXD := 0;
EEDataBuffer.InvertRXD := 0;
EEDataBuffer.InvertRTS := 0;
EEDataBuffer.InvertCTS := 0;
EEDataBuffer.InvertDTR := 0;
EEDataBuffer.InvertDSR := 0;
EEDataBuffer.InvertDCD := 0;
EEDataBuffer.InvertRI := 0;
EEDataBuffer.Cbus0 := 0;
EEDataBuffer.Cbus1 := 0;
EEDataBuffer.Cbus2 := 0;
EEDataBuffer.Cbus3 := 0;
EEDataBuffer.Cbus4 := 0;
EEDataBuffer.RIsVCP := 0;
Result := FT_EE_Read(FT_Handle,@EEDataBuffer);
If Result <> FT_OK then FT_Error_Report('FT_EE_Read ',Result);
end;


function USB_FT_EE_Program : FT_Result;
begin
Result := FT_EE_Program(FT_Handle, @EEDataBuffer);
If Result <> FT_OK then FT_Error_Report('FT_EE_Read ',Result);
end;


function USB_FT_WriteEE(WordAddr:Dword; WordData:Word) : FT_Result;
begin
Result := FT_WriteEE(FT_Handle,WordAddr,WordData);
end;


function USB_FT_ReadEE(WordAddr:Dword) : FT_Result;
begin
Result := FT_ReadEE(FT_Handle,WordAddr,@WordRead);
end;


function USB_FT_EraseEE : FT_Result;
begin
Result := FT_EraseEE(FT_Handle);
end;


function USB_FT_EE_UARead : FT_Result;
begin
Result := FT_EE_UARead(FT_Handle,@UserData,64,@FT_UA_Size);
If Result <> FT_OK then FT_Error_Report('FT_EE_UARead ',Result);
```

end;

function USB_FT_EE_UAWrite : FT_Result;
begin
Result := FT_EE_UAWrite(FT_Handle,@UserData,FT_UA_Size);
If Result <> FT_OK then FT_Error_Report('FT_EE_UAWrite ',Result);
end;

function USB_FT_EE_UASize : FT_Result;
begin
Result := FT_EE_UASize(FT_Handle,@FT_UA_Size);
If Result <> FT_OK then FT_Error_Report('FT_EE_UASize ',Result);
end;

Function Get_USB_Device_LatencyTimer : FT_Result;
Begin
Result := FT_GetLatencyTimer(FT_Handle,@FT_LatencyRd);
If Result <> FT_OK then FT_Error_Report('FT_GetLatencyTimer ',Result);
End;

Function Set_USB_Device_LatencyTimer(Latency:Byte) : FT_Result;
Begin
Result := FT_SetLatencyTimer(FT_Handle, Latency);
If Result <> FT_OK then FT_Error_Report('FT_SetLatencyTimer ',Result);
End;

Function Get_USB_Device_BitMode(var BitMode:Byte) : FT_Result;
Begin
Result := FT_GetBitMode(FT_Handle,@BitMode);
If Result <> FT_OK then FT_Error_Report('FT_GetBitMode ',Result);
End;

Function Set_USB_Device_BitMode(Mask,Enable:Byte) : FT_Result ;
Begin
Result := FT_SetBitMode(FT_Handle,Mask,Enable);
If Result <> FT_OK then FT_Error_Report('FT_SetBitMode ',Result);
End;

Function Set_USB_Parameters(InSize,OutSize:Dword) : FT_Result ;
Begin
Result := FT_SetUSBParameters(FT_Handle,InSize,OutSize);
If Result <> FT_OK then FT_Error_Report('FT_SetUSBParameters ',Result);
End;

End.

unit PORTCONTROLLERMODLib_TLB;

{ This file contains pascal declarations imported from a type library.
This file will be written during each import or refresh of the type
library editor. Changes to this file will be discarded during the
refresh process. }

{ PortController 2.01 }
{ Version 1.0 }

interface

uses Windows, ActiveX, Classes, Graphics, OleCtrls, StdVCL;

const
  LIBID_PORTCONTROLLERMODLib: TGUID = '{30D42D18-9758-4A6B-A4EB-275D1AD4B4D2}';

const

{ Component class GUIDs }
  Class_PortController: TGUID = '{3F40E581-C9DA-42A6-8954-4B70CC916A8D}';

type

{ Forward declarations: Interfaces }
  _IPortControllerEvents = dispinterface;
  IPortController = interface;
  IPortControllerDisp = dispinterface;

{ Forward declarations: CoClasses }
  PortController = IPortController;

{ _IPortControllerEvents Interface }

  _IPortControllerEvents = dispinterface
    ['{12A19D6E-9F84-48F5-82C5-62BD89830A31}']
    procedure DataReceived; dispid 1;
    procedure CtsToggle(NewVal: Smallint); dispid 2;
    procedure DsrToggle(NewVal: Smallint); dispid 3;
    procedure CdToggle(NewVal: Smallint); dispid 4;
    procedure Error; dispid 5;
    procedure Ring; dispid 6;
    procedure BreakSignal; dispid 7;
    procedure TQEmpty; dispid 8;
    procedure EvtCharReceived(const ReadBuffer: WideString; NumBytesRead: Integer);
dispid 9;
  end;

{ IPortController Interface }

  IPortController = interface(IDispatch)
    ['{FCFA025A-DBD7-44B9-8C55-0C4071C874F2}']
    function Get_BaudRate: Integer; safecall;
    procedure Set_BaudRate(Value: Integer); safecall;
    function Get_DataBits: Smallint; safecall;

```
    procedure Set_DataBits(Value: Smallint); safecall;
    function Get_Parity: Smallint; safecall;
    procedure Set_Parity(Value: Smallint); safecall;
    function Get_StopBits: Smallint; safecall;
    procedure Set_StopBits(Value: Smallint); safecall;
    function Get_PortName: WideString; safecall;
    function Get_Cd: Integer; safecall;
    function Get_Cts: Integer; safecall;
    function Get_Dsr: Integer; safecall;
    function Get_Dtr: Integer; safecall;
    procedure Set_Dtr(Value: Integer); safecall;
    function Get_DtrDsr: Integer; safecall;
    procedure Set_DtrDsr(Value: Integer); safecall;
    function Get_RtsCts: Integer; safecall;
    procedure Set_RtsCts(Value: Integer); safecall;
    function Get_Rts: Integer; safecall;
    procedure Set_Rts(Value: Integer); safecall;
    function Get_XonXoff: Integer; safecall;
    procedure Set_XonXoff(Value: Integer); safecall;
    function Get_PortHandle: Integer; safecall;
    function Get_Break: Integer; safecall;
    procedure Set_Break(Value: Integer); safecall;
    function Get_BytesUsedRQ: Integer; safecall;
    function Get_BytesUsedTQ: Integer; safecall;
    function Get_Ring: Integer; safecall;
    function Get_EventChar: WideString; safecall;
    procedure Set_EventChar(const Value: WideString); safecall;
    function Get_FireOnEventChar: Integer; safecall;
    procedure Set_FireOnEventChar(Value: Integer); safecall;
    function Get_EnableReadOnEventChar: Integer; safecall;
    procedure Set_EnableReadOnEventChar(Value: Integer); safecall;
    procedure Open(const PortName, Settings: WideString); safecall;
    procedure Close; safecall;
    function Write(const WriteBuffer: WideString; NumBytesToWrite, Timeout: Integer): Integer; safecall;
    function Read(NumBytesToRead, Timeout: Integer; out pNumBytesRead: Integer): WideString; safecall;
    procedure ClearRQ; safecall;
    procedure ClearTQ; safecall;
    procedure GetErrorStatus(out ParityError, FramingError, OverrunError: Integer); safecall;
    procedure SendXon; safecall;
    procedure SendXoff; safecall;
    function WriteBinary(pWriteBuffer, NumBytesToWrite, Timeout: Integer): Integer; safecall;
    function ReadBinary(NumBytesToRead, Timeout: Integer; out pNumBytesRead: Integer): Integer; safecall;
    function Get_IsOpen: Integer; safecall;
    property BaudRate: Integer read Get_BaudRate write Set_BaudRate;
    property DataBits: Smallint read Get_DataBits write Set_DataBits;
    property Parity: Smallint read Get_Parity write Set_Parity;
    property StopBits: Smallint read Get_StopBits write Set_StopBits;
    property PortName: WideString read Get_PortName;
    property Cd: Integer read Get_Cd;
    property Cts: Integer read Get_Cts;
    property Dsr: Integer read Get_Dsr;
    property Dtr: Integer read Get_Dtr write Set_Dtr;
```

```
    property DtrDsr: Integer read Get_DtrDsr write Set_DtrDsr;
    property RtsCts: Integer read Get_RtsCts write Set_RtsCts;
    property Rts: Integer read Get_Rts write Set_Rts;
    property XonXoff: Integer read Get_XonXoff write Set_XonXoff;
    property PortHandle: Integer read Get_PortHandle;
    property Break: Integer read Get_Break write Set_Break;
    property BytesUsedRQ: Integer read Get_BytesUsedRQ;
    property BytesUsedTQ: Integer read Get_BytesUsedTQ;
    property Ring: Integer read Get_Ring;
    property EventChar: WideString read Get_EventChar write Set_EventChar;
    property FireOnEventChar: Integer read Get_FireOnEventChar write
Set_FireOnEventChar;
    property EnableReadOnEventChar: Integer read Get_EnableReadOnEventChar write
Set_EnableReadOnEventChar;
    property IsOpen: Integer read Get_IsOpen;
  end;

  { DispInterface declaration for Dual Interface IPortController }

  IPortControllerDisp = dispinterface
    ['{FCFA025A-DBD7-44B9-8C55-0C4071C874F2}']
    property BaudRate: Integer dispid 1;
    property DataBits: Smallint dispid 2;
    property Parity: Smallint dispid 3;
    property StopBits: Smallint dispid 4;
    property PortName: WideString readonly dispid 5;
    property Cd: Integer readonly dispid 6;
    property Cts: Integer readonly dispid 7;
    property Dsr: Integer readonly dispid 8;
    property Dtr: Integer dispid 9;
    property DtrDsr: Integer dispid 10;
    property RtsCts: Integer dispid 11;
    property Rts: Integer dispid 12;
    property XonXoff: Integer dispid 13;
    property PortHandle: Integer readonly dispid 14;
    property Break: Integer dispid 15;
    property BytesUsedRQ: Integer readonly dispid 16;
    property BytesUsedTQ: Integer readonly dispid 17;
    property Ring: Integer readonly dispid 18;
    property EventChar: WideString dispid 19;
    property FireOnEventChar: Integer dispid 20;
    property EnableReadOnEventChar: Integer dispid 21;
    procedure Open(const PortName, Settings: WideString); dispid 22;
    procedure Close; dispid 23;
    function Write(const WriteBuffer: WideString; NumBytesToWrite, Timeout: Integer):
Integer; dispid 24;
    function Read(NumBytesToRead, Timeout: Integer; out pNumBytesRead: Integer):
WideString; dispid 25;
    procedure ClearRQ; dispid 26;
    procedure ClearTQ; dispid 27;
    procedure GetErrorStatus(out ParityError, FramingError, OverrunError: Integer); dispid 28;
    procedure SendXon; dispid 29;
    procedure SendXoff; dispid 30;
    function WriteBinary(pWriteBuffer, NumBytesToWrite, Timeout: Integer): Integer; dispid
31;
```

```
    function ReadBinary(NumBytesToRead, Timeout: Integer; out pNumBytesRead: Integer):
Integer; dispid 32;
    property IsOpen: Integer readonly dispid 33;
  end;

{ PortController Class }

  TPortControllerCtsToggle = procedure(Sender: TObject; NewVal: Smallint) of object;
  TPortControllerDsrToggle = procedure(Sender: TObject; NewVal: Smallint) of object;
  TPortControllerCdToggle = procedure(Sender: TObject; NewVal: Smallint) of object;
  TPortControllerEvtCharReceived = procedure(Sender: TObject; const ReadBuffer:
WideString; NumBytesRead: Integer) of object;

  TPortController = class(TOleControl)
  private
    FOnDataReceived: TNotifyEvent;
    FOnCtsToggle: TPortControllerCtsToggle;
    FOnDsrToggle: TPortControllerDsrToggle;
    FOnCdToggle: TPortControllerCdToggle;
    FOnError: TNotifyEvent;
    FOnRing: TNotifyEvent;
    FOnBreakSignal: TNotifyEvent;
    FOnTQEmpty: TNotifyEvent;
    FOnEvtCharReceived: TPortControllerEvtCharReceived;
    FIntf: IPortController;
    function GetControlInterface: IPortController;
  protected
    procedure CreateControl;
    procedure InitControlData; override;
    function GetTOleEnumProp(Index: Integer): TOleEnum;
    procedure SetTOleEnumProp(Index: Integer; Value: TOleEnum);
  public
    procedure Open(const PortName, Settings: WideString);
    procedure Close;
    function Write(const WriteBuffer: WideString; NumBytesToWrite, Timeout: Integer):
Integer;
    function Read(NumBytesToRead, Timeout: Integer; out pNumBytesRead: Integer):
WideString;
    procedure ClearRQ;
    procedure ClearTQ;
    procedure GetErrorStatus(out ParityError, FramingError, OverrunError: Integer);
    procedure SendXon;
    procedure SendXoff;
    function WriteBinary(pWriteBuffer, NumBytesToWrite, Timeout: Integer): Integer;
    function ReadBinary(NumBytesToRead, Timeout: Integer; out pNumBytesRead: Integer):
Integer;
    property ControlInterface: IPortController read GetControlInterface;
    property PortName: WideString index 5 read GetWideStringProp;
    property Cd: Integer index 6 read GetIntegerProp;
    property Cts: Integer index 7 read GetIntegerProp;
    property Dsr: Integer index 8 read GetIntegerProp;
    property PortHandle: Integer index 14 read GetIntegerProp;
    property BytesUsedRQ: Integer index 16 read GetIntegerProp;
    property BytesUsedTQ: Integer index 17 read GetIntegerProp;
    property IPortController_Ring: Integer index 18 read GetIntegerProp;
    property IsOpen: Integer index 33 read GetIntegerProp;
```

published
    property BaudRate: Integer index 1 read GetIntegerProp write SetIntegerProp stored False;
    property DataBits: Smallint index 2 read GetSmallintProp write SetSmallintProp stored False;
    property Parity: Smallint index 3 read GetSmallintProp write SetSmallintProp stored False;
    property StopBits: Smallint index 4 read GetSmallintProp write SetSmallintProp stored False;
    property Dtr: Integer index 9 read GetIntegerProp write SetIntegerProp stored False;
    property DtrDsr: Integer index 10 read GetIntegerProp write SetIntegerProp stored False;
    property RtsCts: Integer index 11 read GetIntegerProp write SetIntegerProp stored False;
    property Rts: Integer index 12 read GetIntegerProp write SetIntegerProp stored False;
    property XonXoff: Integer index 13 read GetIntegerProp write SetIntegerProp stored False;
    property Break: Integer index 15 read GetIntegerProp write SetIntegerProp stored False;
    property EventChar: WideString index 19 read GetWideStringProp write SetWideStringProp stored False;
    property FireOnEventChar: Integer index 20 read GetIntegerProp write SetIntegerProp stored False;
    property EnableReadOnEventChar: Integer index 21 read GetIntegerProp write SetIntegerProp stored False;
    property OnDataReceived: TNotifyEvent read FOnDataReceived write FOnDataReceived;
    property OnCtsToggle: TPortControllerCtsToggle read FOnCtsToggle write FOnCtsToggle;
    property OnDsrToggle: TPortControllerDsrToggle read FOnDsrToggle write FOnDsrToggle;
    property OnCdToggle: TPortControllerCdToggle read FOnCdToggle write FOnCdToggle;
    property OnError: TNotifyEvent read FOnError write FOnError;
    property OnRing: TNotifyEvent read FOnRing write FOnRing;
    property OnBreakSignal: TNotifyEvent read FOnBreakSignal write FOnBreakSignal;
    property OnTQEmpty: TNotifyEvent read FOnTQEmpty write FOnTQEmpty;
    property OnEvtCharReceived: TPortControllerEvtCharReceived read FOnEvtCharReceived write FOnEvtCharReceived;
  end;

procedure Register;

implementation

uses ComObj;

procedure TPortController.InitControlData;
const
  CEventDispIDs: array[0..8] of Integer = (
    $00000001, $00000002, $00000003, $00000004, $00000005, $00000006,
    $00000007, $00000008, $00000009);
  CControlData: TControlData = (
    ClassID: '{3F40E581-C9DA-42A6-8954-4B70CC916A8D}';
    EventIID: '{12A19D6E-9F84-48F5-82C5-62BD89830A31}';
    EventCount: 9;
    EventDispIDs: @CEventDispIDs;
    LicenseKey: nil;
    Flags: $00000000;
    Version: 300);
begin
  ControlData := @CControlData;
end;
```
```

```
procedure TPortController.CreateControl;

  procedure DoCreate;
  begin
    FIntf := IUnknown(OleObject) as IPortController;
  end;

begin
  if FIntf = nil then DoCreate;
end;

function TPortController.GetControlInterface: IPortController;
begin
  CreateControl;
  Result := FIntf;
end;

function TPortController.GetTOleEnumProp(Index: Integer): TOleEnum;
begin
  Result := GetIntegerProp(Index);
end;

procedure TPortController.SetTOleEnumProp(Index: Integer; Value: TOleEnum);
begin
  SetIntegerProp(Index, Value);
end;

procedure TPortController.Open(const PortName, Settings: WideString);
begin
  CreateControl;
  FIntf.Open(PortName, Settings);
end;

procedure TPortController.Close;
begin
  CreateControl;
  FIntf.Close;
end;

function TPortController.Write(const WriteBuffer: WideString; NumBytesToWrite, Timeout: Integer): Integer;
begin
  CreateControl;
  Result := FIntf.Write(WriteBuffer, NumBytesToWrite, Timeout);
end;

function TPortController.Read(NumBytesToRead, Timeout: Integer; out pNumBytesRead: Integer): Widestring;
begin
  CreateControl;
  Result := FIntf.Read(NumBytesToRead, Timeout, pNumBytesRead);
end;

procedure TPortController.ClearRQ;
begin
  CreateControl;
```

```pascal
  FIntf.ClearRQ;
end;

procedure TPortController.ClearTQ;
begin
  CreateControl;
  FIntf.ClearTQ;
end;

procedure TPortController.GetErrorStatus(out ParityError, FramingError, OverrunError: Integer);
begin
  CreateControl;
  FIntf.GetErrorStatus(ParityError, FramingError, OverrunError);
end;

procedure TPortController.SendXon;
begin
  CreateControl;
  FIntf.SendXon;
end;

procedure TPortController.SendXoff;
begin
  CreateControl;
  FIntf.SendXoff;
end;

function TPortController.WriteBinary(pWriteBuffer, NumBytesToWrite, Timeout: Integer): Integer;
begin
  CreateControl;
  Result := FIntf.WriteBinary(pWriteBuffer, NumBytesToWrite, Timeout);
end;

function TPortController.ReadBinary(NumBytesToRead, Timeout: Integer; out pNumBytesRead: Integer): Integer;
begin
  CreateControl;
  Result := FIntf.ReadBinary(NumBytesToRead, Timeout, pNumBytesRead);
end;


procedure Register;
begin
  RegisterComponents('ActiveX', [TPortController]);
end;

end.


unit SPCDemoUnit;

interface

uses
```

Windows, Messages, SysUtils, Variants, Classes, Graphics, Controls, Forms,
Dialogs, StdCtrls, ftSPCClass, Types;

const

mbCaption = 'SPC Demo';

type
  TFormTest = class(TForm)
    cbPorts: TComboBox;
    Label1: TLabel;
    cbBaudRate: TComboBox;
    cbParity: TComboBox;
    cbDataBits: TComboBox;
    Label2: TLabel;
    Label3: TLabel;
    Label4: TLabel;
    cbStopBits: TComboBox;
    cbFlowControl: TComboBox;
    Label5: TLabel;
    Label6: TLabel;
    BtOpen: TButton;
    btClose: TButton;
    checkDCD: TCheckBox;
    checkCTS: TCheckBox;
    checkDSR: TCheckBox;
    checkDTR: TCheckBox;
    checkRING: TCheckBox;
    checkRTS: TCheckBox;
    Memo_Terminal: TMemo;
    FTSPCControl1: FTSPCControl;
    GroupBox1: TGroupBox;
    Label7: TLabel;
    Label8: TLabel;
    Memo_Log: TMemo;
    procedure FormShow(Sender: TObject);
    procedure BtOpenClick(Sender: TObject);
    procedure checkDTRClick(Sender: TObject);
    procedure checkRTSClick(Sender: TObject);
    procedure btCloseClick(Sender: TObject);
    procedure cbPortsChange(Sender: TObject);
    procedure cbBaudRateChange(Sender: TObject);
    procedure cbParityChange(Sender: TObject);
    procedure cbDataBitsChange(Sender: TObject);
    procedure cbStopBitsChange(Sender: TObject);
    procedure cbFlowControlChange(Sender: TObject);
    procedure Memo_TerminalKeyDown(Sender: TObject; var Key: Word;
      Shift: TShiftState);
    procedure Memo_TerminalKeyPress(Sender: TObject; var Key: Char);
    procedure FTSPCControl1Ring(Sender: TObject);
    procedure FTSPCControl1Dsr(Sender: TObject);
    procedure FTSPCControl1Cts(Sender: TObject);
    procedure FTSPCControl1Dcd(Sender: TObject);
    procedure FTSPCControl1Receive(Sender: TObject; Count: Cardinal);
    procedure FTSPCControl1Error(Sender: TObject; Error: FTSPCCommErrors);
    procedure FTSPCControl1EventChar(Sender: TObject);

```
    procedure FTSPCControl1Break(Sender: TObject);
  private
    { Private declarations }
  public
    { Public declarations }
  end;

  function BoolToStr_(B: Boolean):string;

var
  FormTest: TFormTest;

implementation

{$R *.dfm}

function BoolToStr_(B: Boolean):string;
begin
  if B then result := 'True'
   else result := 'False';
end;

procedure TFormTest.FormShow(Sender: TObject);
var
  i: integer;
  portsCount: word;
begin
  try
    portsCount := FTSPCControl1.EnumSerialPorts;
    if portsCount > 0 then
      begin
        cbPorts.Items.Clear;
        for i := 0  to portsCount-1 do
          begin
            cbPorts.Items.Add(FTSPCControl1.GetSerialPort(i));
          end;
        cbPorts.ItemIndex := 0;
        btOpen.Enabled := true;
      end
    else
      Application.MessageBox('System have no any serial port.',
       mbCaption ,MB_OK+MB_ICONWARNING)
  except
    on E:FTSPCException do
      Application.MessageBox(PChar('Error ' + inttostr(E.ErrorCode) + #13#10 +
        E.SystemErrorMessage + #13#10 + E.ErrorSource), mbCaption
,MB_OK+MB_ICONERROR);
   end;
end;

procedure TFormTest.BtOpenClick(Sender: TObject);
begin
  try
    FTSPCControl1.PortName := cbPorts.Text;
    // setting BaudRate
    FTSPCControl1.BaudRate := strtoint(cbBaudRate.Text);
```

```pascal
    // setting Parity
    FTSPCControl1.Parity := FTSPCParity(cbParity.ItemIndex);
    // setting ByteSize
    FTSPCControl1.DataBits := FTSPCDataBits(cbDataBits.ItemIndex);
    // setting StopBits
    FTSPCControl1.StopBits := FTSPCStopBits(cbStopBits.ItemIndex);
    // setting FlowControll
    FTSPCControl1.FlowControl := FTSPCFlowControl(cbFlowControl.ItemIndex);
    // open port
    FTSPCControl1.Open();

    Memo_Log.Lines.add('Port ' + FTSPCControl1.PortName + ' opened.');
    Memo_Terminal.ReadOnly := false;

    btOpen.Enabled := False;
    btClose.Enabled := True;
    cbPorts.Enabled := False;

    CheckDTR.Enabled := True;
    CheckRTS.Enabled := True;

    if FTSPCControl1.Dtr then
      CheckDTR.Checked := true
    else
      CheckDTR.Checked := false;

    if FTSPCControl1.Rts then
      CheckRTS.Checked := true
    else
      CheckRTS.Checked := false;

  except
    on E:FTSPCException do
      Application.MessageBox(PChar('Error ' + inttostr(E.ErrorCode) + #13#10 +
        E.SystemErrorMessage + #13#10 + E.ErrorSource), mbCaption
,MB_OK+MB_ICONERROR);
  end;
end;

procedure TFormTest.checkDTRClick(Sender: TObject);
begin
  try
    if CheckDTR.Checked then FTSPCControl1.Dtr := True
    else FTSPCControl1.Dtr := False;
  except
    on E:FTSPCException do
      Application.MessageBox(PChar('Error ' + inttostr(E.ErrorCode) + #13#10 +
        E.SystemErrorMessage + #13#10 + E.ErrorSource), mbCaption
,MB_OK+MB_ICONERROR);
  end;
end;

procedure TFormTest.checkRTSClick(Sender: TObject);
begin
  try
    if CheckRTS.Checked then FTSPCControl1.Rts := True
```

```
    else FTSPCControl1.Rts := False;
  except
    on E:FTSPCException do
      Application.MessageBox(PChar('Error ' + inttostr(E.ErrorCode) + #13#10 +
        E.SystemErrorMessage + #13#10 + E.ErrorSource), mbCaption
,MB_OK+MB_ICONERROR);
  end;
end;

procedure TFormTest.btCloseClick(Sender: TObject);
begin
  FTSPCControl1.Close;

  Memo_Log.Lines.add('Port ' + FTSPCControl1.PortName + ' closed.');

  Memo_Terminal.ReadOnly := true;

  btOpen.Enabled := True;
  btClose.Enabled := False;

  CheckDTR.Enabled := False;
  CheckRTS.Enabled := False;
  FTSPCControl1.Rts := False;
  FTSPCControl1.Dtr := False;

  cbPorts.Enabled := True;
  checkDCD.Checked := false;
  checkCTS.Checked := false;
  checkDSR.Checked := false;
  checkDTR.Checked := false;
  checkRING.Checked := false;
  checkRTS.Checked := false;
end;

procedure TFormTest.cbPortsChange(Sender: TObject);
begin
  if FTSPCControl1.IsOpened then
    begin
      btClose.Click; // FTSPCControl1.Close;
      btOpen.Click; // FTSPCControl1.Open();
    end;
end;

procedure TFormTest.cbBaudRateChange(Sender: TObject);
begin
  try
    FTSPCControl1.BaudRate := strtoint(cbBaudRate.Text);
  except
    on E:FTSPCException do
      Application.MessageBox(PChar('Error ' + inttostr(E.ErrorCode) + #13#10 +
        E.SystemErrorMessage + #13#10 + E.ErrorSource), mbCaption
,MB_OK+MB_ICONERROR);
  end;
end;

procedure TFormTest.cbParityChange(Sender: TObject);
```

```
begin
 try
   FTSPCControl1.Parity := FTSPCParity(cbParity.ItemIndex);
 except
   on E:FTSPCException do
     Application.MessageBox(PChar('Error ' + inttostr(E.ErrorCode) + #13#10 +
       E.SystemErrorMessage + #13#10 + E.ErrorSource), mbCaption
,MB_OK+MB_ICONERROR);
 end;
end;

procedure TFormTest.cbDataBitsChange(Sender: TObject);
begin
 try
   FTSPCControl1.DataBits := FTSPCDataBits(cbDataBits.ItemIndex);
 except
   on E:FTSPCException do
     Application.MessageBox(PChar('Error ' + inttostr(E.ErrorCode) + #13#10 +
       E.SystemErrorMessage + #13#10 + E.ErrorSource), mbCaption
,MB_OK+MB_ICONERROR);
 end;
end;

procedure TFormTest.cbStopBitsChange(Sender: TObject);
begin
 try
   FTSPCControl1.StopBits := FTSPCStopBits(cbStopBits.ItemIndex);
 except
   on E:FTSPCException do
     Application.MessageBox(PChar('Error ' + inttostr(E.ErrorCode) + #13#10 +
       E.SystemErrorMessage + #13#10 + E.ErrorSource), mbCaption
,MB_OK+MB_ICONERROR);
 end;
end;

procedure TFormTest.cbFlowControlChange(Sender: TObject);
begin
 try
   FTSPCControl1.FlowControl := FTSPCFlowControl(cbFlowControl.ItemIndex);
 except
   on E:FTSPCException do
     Application.MessageBox(PChar('Error ' + inttostr(E.ErrorCode) + #13#10 +
       E.SystemErrorMessage + #13#10 + E.ErrorSource), mbCaption
,MB_OK+MB_ICONERROR);
 end;
end;

procedure TFormTest.Memo_TerminalKeyDown(Sender: TObject; var Key: Word;
 Shift: TShiftState);
begin
// Key contains enhanced Virtual key codes
// They not always correspond to table of ASCII codes.
// Therefore event OnKeyPress is used.
end;

procedure TFormTest.Memo_TerminalKeyPress(Sender: TObject; var Key: Char);
```

```
const
  CRLF: array[0..1] of byte = ($0D,$0A);
begin
 if FTSPCControl1.IsOpened then
  begin
   try
    if key = #13 then
     begin
      FTSPCControl1.Write(@CRLF[0], 2);
     end
    else
     FTSPCControl1.Write(@Key,1);
   except
    on E:FTSPCException do
     Application.MessageBox(PChar('Error ' + inttostr(E.ErrorCode) + #13#10 +
      E.SystemErrorMessage + #13#10 + E.ErrorSource), mbCaption
,MB_OK+MB_ICONERROR);
   end;
  end;
end;

procedure TFormTest.FTSPCControl1Ring(Sender: TObject);
var
  State: Boolean;
begin
 State := FTSPCControl1.Ring;
 Memo_Log.Lines.add('Port event -> change RING state (' + BoolToStr_(State) + ')');
 if State then CheckRING.Checked := true
   else CheckRING.Checked := False;
end;

procedure TFormTest.FTSPCControl1Dsr(Sender: TObject);
var
  State: Boolean;
begin
 State := FTSPCControl1.Dsr;
 Memo_Log.Lines.add('Port event -> change DSR state (' + BoolToStr_(State) + ')');
 if State then CheckDSR.Checked := true
   else CheckDSR.Checked := False;
end;

procedure TFormTest.FTSPCControl1Cts(Sender: TObject);
var
  State: Boolean;
begin
 State := FTSPCControl1.Cts;
 Memo_Log.Lines.add('Port event -> change CTS state (' + BoolToStr_(State) + ')');
 if State then CheckCTS.Checked := true
   else CheckCTS.Checked := False;
end;

procedure TFormTest.FTSPCControl1Dcd(Sender: TObject);
var
  State: Boolean;
begin
 State := FTSPCControl1.Dcd;
```

```
    Memo_Log.Lines.add('Port event -> change DCD state (' + BoolToStr_(State) + ')');
    if State then CheckDCD.Checked := true
      else CheckDCD.Checked := False;
end;

procedure TFormTest.FTSPCControl1Receive(Sender: TObject; Count: Cardinal);
var
  inBuffer: array of byte;
  ReadCnt: integer;
  i: integer;
begin
  try
    SetLength(inBuffer, Count);
    ReadCnt := FTSPCControl1.ReadArray(inBuffer);
    if ReadCnt > 0 then
      for i := 0 to ReadCnt-1 do
        begin
          Memo_Terminal.Text := Memo_Terminal.Text + char(inBuffer[i]);
        end;
  except
    on E:FTSPCException do
      Application.MessageBox(PChar('Error ' + inttostr(E.ErrorCode) + #13#10 +
      E.SystemErrorMessage + #13#10 + E.ErrorSource), mbCaption
,MB_OK+MB_ICONERROR);
  end;
end;

procedure TFormTest.FTSPCControl1Error(Sender: TObject;
  Error: FTSPCCommErrors);
begin
  case Error of
    ftspcErrorFrame: Memo_Log.Lines.add('event -> error was occured (ftspcErrorFraming)');
    ftspcErrorParity: Memo_Log.Lines.add('event -> error was occured (ftspcErrorParity)');
    ftspcErrorOverrun: Memo_Log.Lines.add('event -> error was occured
(ftspcErrorOverrun)');
  end;
end;

procedure TFormTest.FTSPCControl1EventChar(Sender: TObject);
begin
  Memo_Log.Lines.add('event -> received EVENT character.');
end;

procedure TFormTest.FTSPCControl1Break(Sender: TObject);
begin
  Memo_Log.Lines.add('event -> received BREAK signal.');
end;

end.
```

10.1 References

http://www.ftdichip.com/
FTDI USB DLL and VCP convertors
http://www.farnell.com/
Worldwide component supplier including PIC's and FTDI and 4D Systems USB converters
http://microchip
PIC datasheets and MPLAB software
http://www.sparkfun.com/
Electronic modules and components
http://www.active-robots.com/
Electronic compass module
http://www.elektor-electronics.co.uk/
International Electronics Magazine
http://www.x-ways.net/
WINHEX
http://www.nirsoft.net/
USBDVIEW
http://www.mpfreezone.com/
Terminal v2.14
http://www.scientificcomponent.com/
PortController Active X
http://www.lohninger.com/sdlindex.html
SDL Component suite
http://www.cstsoft.com/english/downloads.htm
CST vitual Instument components
http://www.iocomp.com/Downloads/Evaluations.aspx
IOCOM virtual Instument components
http://www.abaecker.biz/
Abacus Delphi components
http://www.uniworkstech.com/download.htm
Uniworks vitual instrument components
http://www.tmssoftware.com/site/tiw.asp
TMS instrument workshop
http://www.4dsystems.com.au/prod.php?id=18
Micro USB VCP bridge
http://www.componentsource.com/products/global-majic-aircraft
GMS Active X virtual flight deck components
http://delphi.about.com/od/vclwriteenhance/l/aa061104a.htm
VCL Tcolour button component

http://www.delphibasics.co.uk/Article.asp?Name=FirstPgm
Delphi tutorial
http://www.codegear.com/downloads/free/delphi
Borland Delphi
http://www.prolific.com.tw/eng/downloads.asp?ID=31
Prolific USB VCP bridge drivers
http://www.maplin.co.uk
N42FL function generator
http://www.virtins.com
virtual multi instrument
http://www.dataq.com/products/software/xcontrols.htm
Active X controls
http://www.usbnow.co.uk/
BELKIN USB RS232 adapters
http://www.prolific.com.tw/eng/downloads.asp?ID=31
Belkin and prolific USB RS232 driver downloads
http://www.brothersoft.com/delphi-76437.html
Delphi 7 download

10.2 INDEX

1

12F629, 106

4

44 pin demo board, 95, 118
4D, 23

7

74HC573, 101, 107, 109

A

Abacus, 1, 56, 59, 93, 259
ABACUS, 157
AbBinary, 59
Abthermometer, 93
Active X, 1, 2, 4, 63, 77, 80, 185, 186, 259, 260
ADC, 2, 56, 80, 118, 134, 135, 137, 163, 165, 170, 174, 177, 178
aircraft, 185, 192, 203, 259
analogue, 56, 57, 84, 93, 94, 134, 186
ANGULAR.PAS, 210
Anibutton.pas, 211
Appendix, 2, 77, 95, 225
ASCII, 33, 34, 51, 80, 82, 256
asynchronous, 4, 14, 29, 47

B

bin number, 177
bit-bang, 1, 29, 118
button, 1, 56, 62, 78, 88, 113, 141, 206, 259

C

car, 56, 142, 144, 185, 186, 187, 188, 189
CMPS03, 151, 153
colorbutton, 88, 113
COM, 4, 63, 69, 71, 72, 77, 88, 94, 95, 127, 134, 199, 200, 201, 202
compass, 2, 3, 150, 151, 160, 161, 162, 221, 259
Compass, 2, 151, 157, 160, 161, 162, 219, 220
Compasstest, 162
COMPASSX2.PAS, 157
Component.Value, 56, 185, 186
connectors, 74, 76, 100
CST, 2, 222, 223, 224, 259

D

D2XXUnit, 95, 111, 117, 123, 190, 225
DAC, 1, 109, 110, 114, 174
DAC0800N, 109
degrees C, 87, 89
DELPHIFFTX, 180
Device Manager, 70
DLL, 1, 4, 95, 96, 98, 101, 102, 103, 111, 117, 118, 123, 133, 139, 189, 190, 230, 231, 232, 259
DLP-USB245M, 4, 101, 104, 134, 151
DS75, 84, 87
DSO, 133, 134, 137, 138, 139, 141, 142, 143, 144
DSOX280.PAS, 139
DSP, 163, 164, 168, 176, 177, 178, 180
dsPIC30F6012a, 164, 166, 177
DSPX, 174, 176, 177
DTR, 86, 124, 190, 227, 228, 229, 236
D-Type, 73, 74

E

EVAL232R, 4, 18, 23, 27, 28, 63, 64, 75, 118, 119, 121, 189
EXCEL, 3, 86, 89, 90, 141, 180

F

Farnell, 259

FFT, 2, 3, 163, 167, 168, 170, 171, 172, 173, 174, 177, 180, 194
freezer, 88
FT232R, 126, 226, 240, 241, 242, 243
Ftclean, 95
FTClean, 95
FTDI, 1, 2, 4, 63, 86, 95, 96, 100, 101, 110, 118, 119, 126, 127, 133, 139, 164, 174, 189, 259
FTDIFT8U2XX, 98

G

gauge, 187, 194, 195, 202, 203, 204
gender changers, 76
GMS, 2, 185, 186, 187, 189, 192, 259
graph, 89, 94
guage, 1, 59, 118, 122, 125, 195, 219

H

hub, 21
Hyperterminal, 52

I

ICL7660CPA, 134, 135
icon, 79, 130
Index, 2, 227, 230, 232, 239, 248, 250
intervals, 176
IOCOMP, 2, 217, 218

J

joystick, 1, 2, 3, 126, 128, 129, 130, 131

L

Lambda, 2, 142
LAMDA2X, 142
LED, 1, 56, 59, 60, 61, 95, 101, 102, 103, 107, 108, 109, 110, 113, 114, 164, 172, 176, 180, 185, 206, 207, 208

LED test board, 1, 95, 102, 109, 110, 113, 114

M

magnetic declination, 151
Magnify, 140
Maplins, 174
MC1458CPI, 134, 135
microchip, 1, 118, 259
MICROCHIPS 44 PIN DEMO, 46
Microsoft, 19, 130, 131, 150, 163, 199, 201
MM232R, 4, 22, 23, 24, 164, 165, 174, 176, 177, 179
mouse, 1, 2, 3, 56, 57, 62, 126, 127, 128, 129, 130, 131, 132, 203
MOUSE84.ASM, 131

N

NumberxDLL, 110

O

Object inspector, 59, 122
Object Inspector, 6, 189
oscilloscope, 2, 133, 139, 140

P

p16F887, 120
panel, 1, 56, 60, 110, 113, 114, 130, 199
Paneldllx2, 113
peak frequency, 163
PIC, 3, 56, 84, 86, 88, 93, 94, 118, 120, 134, 135, 138, 150, 151, 174, 180, 186, 259
PIC12C508, 48
PIC12F629, 3, 37, 40, 84, 101, 106
PIC16F84, 3, 126, 151
PIC16F84A, 30, 31, 35
PIC16F887, 3, 46, 48
PIC18F452, 43, 45
PIC30F6012A, 163
PICKIT 2, 86
PICLC71-20/P, 133

PortController, 4, 77, 78, 79, 245, 248, 259
POTDLLX, 122
potentiometer, 3, 95, 118, 134, 174, 186
Program Numberxxdll.pas, 110
Program 'Meterdemo.pas', 215
Program A2D.ASM, 120
Program CAR.PAS, 189
Program DelphiFFTX.PAS, 182
Program Guagesample, 195
Program LED1x, 60
Program Rotary.PAS, 212
Program SIMUL.C, 170
program UNIT11.PAS, 116
Project CARX, 187
Project PRJMETER, 193
Project Simul, 169
Project VIX, 81
Prolific, 4, 63, 86, 260
PROLIFIC/BELKIN, 18, 30

R

RS232, 1, 3, 4, 5, 13, 14, 16, 17, 18, 19, 22, 24, 28, 29, 34, 35, 40, 48, 51, 53, 54, 63, 64, 69, 74, 75, 76, 77, 80, 84, 86, 95, 101, 104, 110, 118, 121, 127, 164, 174, 260
RTS, 86, 124, 126, 190, 226, 227, 228, 229, 236

S

SDL, 2, 193, 194, 195, 196, 198, 259
sensor33-2400.ASM, 88
simple_thermometerx, 57
slider, 1, 110, 140, 151
SPBRG, 45, 47
speedometer, 56, 186, 187, 188, 189

Superterm, 73

T

Terminal emulation, 1, 54
thermometer, 1, 3, 56, 57, 63, 84, 88, 89, 91, 93, 94
TMS, 2, 206, 259

U

UART, 4, 163, 172
UB232R, 2, 4, 22, 23, 31, 127
Uniworks, 2, 12, 114, 125, 199, 205, 259
USB, 1, 3, 4, 5, 17, 18, 19, 20, 21, 22, 24, 29, 35, 40, 51, 54, 56, 63, 64, 67, 69, 71, 73, 74, 77, 80, 84, 86, 88, 95, 96, 98, 100, 101, 104, 106, 111, 112, 117, 118, 119, 124, 125, 126, 127, 133, 134, 151, 164, 165, 174, 176, 189, 190, 191, 226, 227, 228, 230, 234, 235, 236, 237, 238, 239, 241, 242, 243, 244, 259, 260
USBADC3.ASM, 138
USBOUT, 95, 102, 106
USBView, 98, 99

V

VCP, 1, 4, 63, 66, 77, 84, 86, 95, 96, 127, 134, 164, 259, 260
vcp_ftdi_temp, 88, 91, 94
vcp_temp.pas, 88

W

wizard, 96